Modern Mathematics Revision

# Modern Mathematics Revision

by

**J. Clay** B.Sc., A.R.C.S. and **A. R. Trippett** M.A.

**London · George Allen & Unwin Ltd**
Ruskin House Museum Street

First published in 1972

© George Allen & Unwin Ltd 1972

ISBN 0 04 510039 X

Printed in Great Britain
in 11 point Times New Roman type
by Cox & Wyman Ltd,
London, Fakenham and Reading

# Contents

|  |  | *page* |
|---|---|---|
| 1 | Sets | 1 |
| 2 | Sets of Numbers | 6 |
| 3 | Implication and Indices | 10 |
| 4 | Matrices | 12 |
| 5 | Relations and Functions | 18 |
| 6 | Translations and Vectors | 23 |
| 7 | Reflection and Rotation | 28 |
| 8 | Enlargement and Shearing | 32 |
| 9 | Transformations and Matrices | 36 |
| 10 | Graphical Representation of Equations | 44 |
| 11 | Equations and Inequalities | 48 |
| 12 | Simultaneous Equations | 52 |
| 13 | Linear Programming | 58 |
| 14 | Trigonometry | 64 |
| 15 | Significance in Numbers and Number Bases | 70 |
| 16 | Computation | 74 |
| 17 | Logarithms | 77 |
| 18 | The Slide Rule | 81 |
| 19 | Formulae | 83 |
| 20 | Proportionality | 86 |
| 21 | Calculations from Graphs | 93 |
| 22 | Probability | 99 |
| 23 | Statistics | 106 |
| 24 | Area and Volume | 113 |
| 25 | The Circle and the Sphere | 117 |
| 26 | Topology | 123 |
| 27 | Geometry | 130 |
| 28 | Structure | 137 |
| | Miscellaneous Exercises | 143 |
| | Answers | 157 |

# Chapter 1
# Sets

A set is a collection of objects. The objects contained in the set are called elements. A set can be defined descriptively, e.g. the set of vowels in the English alphabet, or explicitly, e.g. the set whose elements are $a, e, i, o, u$. Shorthand ways of writing these sets are {vowels in the English alphabet}, $\{a, e, i, o, u\}$. A further way of defining a set is to write it in the form $\{x:x$ satisfies a condition}, which is the set consisting of all the objects satisfying the given condition. The above set, for example, would be $\{x:x$ is a vowel in the English alphabet}.

If $A$ is a set, $c \in A$ means that $c$ is an element of $A$ and $c \notin A$ means that $c$ is not an element of $A$.

For two sets $A$ and $B$ the intersection of $A$ and $B$ (written $A \cap B$) is defined as the set of elements which are members of both $A$ and $B$, i.e.

$$A \cap B = \{x:x \in A \text{ and } x \in B\}.$$

If $A$ and $B$ have no elements in common, their intersection will contain no elements. The set containing no elements, the empty set, is written { } or $\varnothing$. If $A \cap B = \varnothing$, the sets $A$ and $B$ are disjoint.

The union of $A$ and $B$ (written $A \cup B$) is the set of all elements which are elements of $A$ or elements of $B$ or elements of both sets, i.e.

$$A \cup B = \{x:x \in A \text{ or } x \in B \text{ or both}\}.$$

$A$ is a subset of $B$ (written $A \subset B$) if every element of $A$ is an element of $B$.

If a number of sets $A, B, C, \ldots$ are under consideration, the universal set (written $\xi$) for these sets is any set which includes all the elements of $A, B, C, \ldots$ amongst its elements.

1

The complement of a set $A$ (written $A'$) is the set of elements of the universal set which are not elements of $A$, i.e.

$$A' = \{x : x \in \xi \text{ and } x \notin A\}.$$

Venn diagrams offer a convenient pictorial representation of sets and the relationships between them. In Fig. 1.1, the

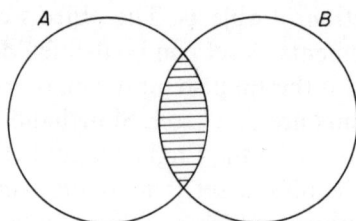

FIG. 1.1

interior of the shape $A$ represents all the elements of $A$ and similarly for $B$. Hence the shaded area represents $A \cap B$. Both the sets $A$ and $B$ must lie inside the universal set. In Fig. 1.2

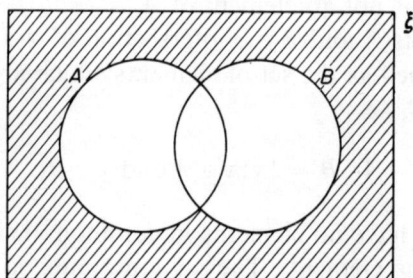

FIG. 1.2

the shaded area represents $(A \cup B)'$, the complement of $A \cup B$. In general the Venn diagram for three sets will be as shown in Fig. 1.3.

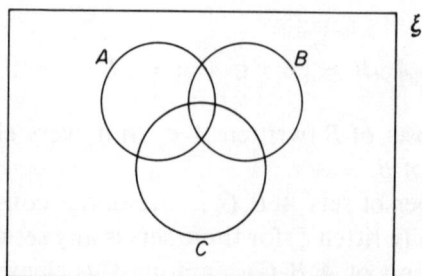

FIG. 1.3

2

# EXERCISES

1. If $\xi = \{a, b, c, d, e, f, g, h, i, j, k, l, m, n,\}$, $A = \{a, c, f, h, i, l\}$ and $B = \{a, d, f, h, j, k\}$, list the elements of the sets

    (a) $A \cap B$,     (b) $A \cup B$,     (c) $A'$,

    (d) $B'$,     (e) $(A \cap B)'$,     (f) $A' \cap B$,

    (g) $A \cup B'$,     (h) $A \cup \xi$,     (i) $B \cap \xi$.

2. If in addition to the sets defined in question 1, $C = \{a, c, h, j, k, m, n\}$, list the elements of the sets

(a) $(A \cap B) \cup C$,     (b) $(A \cup B) \cup C$,   (c) $(A \cup B') \cap C$,

(d) $(A \cap B)' \cap C$,     (e) $(A \cup B) \cap C'$,   (f) $(A \cap C) \cup (B \cap C)$,

(g) $(A \cup C)' \cap (A \cap C')$.

3. Represent each of your answers to question 2 on a Venn diagram. Indicate clearly which part of the diagram corresponds to your answer.

4. Use Venn diagrams to prove that

$$\left.\begin{array}{l} A \cup (B \cup C) = (A \cup B) \cup C \\ A \cap (B \cap C) = (A \cap B) \cap C \end{array}\right\} \quad \text{associative law,}$$

$$\left.\begin{array}{l} A \cap (B \cup C) = (A \cap B) \cup (A \cap C) \\ A \cup (B \cap C) = (A \cup B) \cap (A \cup C) \end{array}\right\} \quad \text{distributive law,}$$

$$\left.\begin{array}{l} (A \cup B)' = A' \cap B' \\ (A \cap B)' = A' \cup B' \end{array}\right\} \quad \text{De Morgan's Laws.}$$

Use Venn diagrams to simplify
(a) $(A \cap B) \cup B$,   (b) $(A \cup B) \cap (B' \cup A)$,   (c) $(A' \cap B) \cap (A \cap B')$.

5. (i) Simplify, assuming that $A$ and $B$ are not empty,

    (a) $A \cup A$,     (b) $A \cap \varnothing$,   (c) $A \cup \xi$,

    (d) $\varnothing \cap \xi$,     (e) $A \cap A'$,   (f) $A' \cap (B \cap A)$.

  (ii) State whether the following are true or false. Modify any incorrect statements to make them true.

      (a) $A \cap B = A \Rightarrow A \subset B$,

      (b) $A \subset B$ and $B \subset A \Rightarrow A = B$,

3

MODERN MATHEMATICS REVISION

(c) $A' \cap B' = \emptyset \Rightarrow A \cup B = \emptyset$,
(d) $A \cap B' = \emptyset \Rightarrow A \subset B$.

6. If $\xi = \{$houses in the U.K.$\}$,
   $A = \{$houses built in the last five years$\}$,
   $B = \{$centrally heated houses$\}$,
   $C = \{$houses with two storeys$\}$,
express the following statements symbolically.

(a) All houses built in the last five years are centrally heated.
(b) No houses with two storeys are centrally heated.
(c) If a two-storeyed house has central heating then it was built in the last five years.
(d) All houses either have central heating, or have two storeys and were built in the last five years.

If (a) and (b) are both true what conclusion can you draw? (Venn diagrams may be helpful.)

7. If $\xi = \{$pupils in a school$\}$,
   $A = \{$pupils in white house$\}$,
   $B = \{$pupils who are good at maths$\}$,
   $C = \{$members of the school chess club$\}$,
express each of the following in words.

(a) $C \subset B$,  (b) $A \cap B \neq \emptyset$,  (c) $B \cup C = \xi$,
(d) $B \cap C \subset A$,  (e) $A' \cap C = \emptyset$,  (f) $C \subset A \cup B'$.

8. In interviewing people about their hobbies it is found that fifteen people collect stamps, five collect coins and six collect match-box labels. Everyone who collects coins also collects stamps. Five people collect stamps and match-box labels, and two people collect all three things. Assuming that everyone interviewed collects at least one of the three things, how many people were interviewed and how many collect stamps only?

9. $n(A) = $ the number of elements in the set $A$. Three sets are such that $n(A) = 8$, $n(B) = 15$, $n(C) = 7$, $n(A \cup B \cup C) = 21$, $n(B \cap C \cap A') = 5$ and $A \cap B \cap C' = A \cap C \cap B' = \emptyset$. Find $n(A \cap B \cap C)$.

4

10. A world-cup soccer squad consists of twenty-two players, each of them able to play in one or more positions of defence, midfield and attack. The number of players who can play only in midfield is twice the number of those who can play in attack and midfield, and two more than the number who can play only in attack. The number of players who can play in defence and midfield is twice the number who can play in midfield and attack, whilst the number who play in defence only is the same as the number who play in attack only. If all players who can defend and attack can also play in midfield, and fourteen can play in only one position, find the number of players who can play in all three positions and the number who can play in attack.

11. Cars taking the M.O.T. test at a certain garage fail on one or more of the following tests: steering, brakes and lights. Three times as many fail on brakes alone as fail on steering and lights. Six fail on both steering and brakes, and twenty fail on steering. The number failing on steering and lights is the same as the number failing on lights alone, and two more than the number failing on brakes and lights (but not on all three). If forty-three fail the test and thirty-one fail on one thing only, how many cars fail on all three?

# Chapter 2
# Sets of Numbers

Elements of the set $Z = \{\dots\ ^-3,\ ^-2,\ ^-1, 0, 1, 2, 3\dots\}$ are called directed numbers or integers. In this set 0 is the additive identity, since

$$x + 0 = 0 + x = x \qquad \text{for all } x \in Z.$$

Each element of $Z$ has an additive inverse, i.e. given any integer $a$ there is another integer $b$ such that $a + b = b + a = 0$, e.g.

the inverse of 3 is $^-3$ since $3 + {}^-3 = {}^-3 + 3 = 0$,
the inverse of $^-7$ is 7 since $^-7 + 7 = 7 + {}^-7 = 0$.

The rules of addition and subtraction of integers are as follows.

(a) *Addition*   Here it is convenient to consider the integers as shift numbers. Thus 3 could represent 'move 3 places to the right' and $^-5$ could represent 'move 5 places to the left'. Addition then corresponds to the operation 'followed by'. Thus $3 + {}^-7 = {}^-4$, since moving 3 places right followed by 7 left corresponds to moving 4 places left.

(b) *Subtraction*   Subtracting an integer is equivalent to adding the additive inverse of that integer. Thus

$$3 - 7 = 3 + {}^-7 = {}^-4$$
$$3 - {}^-7 = 3 + 7 \;\;= 10.$$

The table in Fig. 2.1 shows how to obtain the sign of the answer when multiplying and dividing, but it is vital to remember that the rules only hold in these two cases.

If $a$ and $b$ are natural numbers, i.e. elements of the set $\{1, 2, 3\dots\}$ and $a$ is exactly divisible by $b$, then $b$ is said to be a factor of $a$, e.g. the set of factors of 24 is $\{1, 2, 3, 4, 6, 8, 12, 24\}$.

6

| X or ÷ | + | − |
|---|---|---|
| + | + | − |
| − | − | + |

Fig. 2.1

A natural number, other than 1, is prime if its only factors are 1 and itself. Any number can be expressed as a product of prime factors, e.g.

$$24 = 2 \times 2 \times 2 \times 3 = 2^3 \times 3.$$

In addition to the integers and the natural numbers, other sets of numbers are

(a) {Rational numbers}, i.e. the set of numbers which can be expressed as fractions; or as decimals which terminate or recur, e.g. 7, $^-4\frac{3}{4}$, 3·91203, 7·$\dot{2}$1$\dot{3}$.

(b) {Irrational numbers}, i.e. the set of numbers which when expressed in decimal form do not recur or terminate, and thus cannot be expressed as fractions, e.g. $\sqrt{2}$, $\pi$, $3\sqrt{7}+6$.

Note that {Rational numbers}∩{Irrational numbers} = ∅.

(c) {Real numbers} = {Rational numbers} ∪{Irrational numbers}. We have

{Natural numbers} ⊂ {Integers}
⊂ {Rational numbers} ⊂ {Real numbers}.

## EXERCISES

1. Calculate

(a) $12 + {}^-7$,  (b) $3 + {}^-7$,  (c) $^-3 + {}^-4$,
(d) $12 - 13$,  (e) $^-2 - {}^-3$,  (f) $^-3 \times 4$,

7

(g) $12 \div {}^-3,$　　　　(h) ${}^-7 \times (3 - {}^-2),$　　(i) $({}^-3+2) \times 6,$
(j) ${}^-12 \div (3 - {}^-2),$　　(k) $(2 + {}^-2) \times {}^-6,$　　(l) ${}^-3+7-{}^-6.$

2. Insert in place of the dots the number required to make the following calculations correct.

(a) $13 + \ldots = 8,$　　　　　(b) ${}^-3 - \ldots = {}^-1,$
(c) $\ldots \times {}^-4 = 6,$　　　　(d) $(\ldots - {}^-4) \div {}^-3 = {}^-4,$
(e) ${}^-2 \times (4 + \ldots) = {}^-2,$　　(f) $7 - \ldots + {}^-5 = 5.$

3. List (i) the set of factors (ii) the set of prime factors of

(a) 15, (b) 36, (c) 23, (d) 108, (e) 512.

4. Express 42 and 30 as products of prime factors. Hence, find the largest number which is a factor of both 42 and 30, and the smallest number of which both 42 and 30 are factors. Repeat this question using

(a) 30 and 84, (b) 120 and 132, (c) 29 and 53.

5. Complete the table to show to which of the sets $A = \{\text{natural nos}\}$ $B = \{\text{integers}\}$ $C = \{\text{rational nos}\}$ $D = \{\text{irrational nos}\}$ and $E = \{\text{real nos}\}$ the given numbers belong.

|  | A | B | C | D | E |
|---|---|---|---|---|---|
| $3\frac{3}{4}$ | × | × | √ | × | √ |
| ${}^-5$ |  |  |  |  |  |
| 2·61 |  |  |  |  |  |
| $\sqrt{90}$ |  |  |  |  |  |
| 0·71̇6̇ |  |  |  |  |  |
| $\pi^2$ |  |  |  |  |  |
| $\sqrt{3}$ |  |  |  |  |  |

6. Give the solution sets of each of the following equations if
(a) $x$ is a natural number, (b) $x$ is rational, (c) $x$ is real.

(i) $x + 3 = 1,$　　　　(ii) $3x = 7,$　　　　(iii) $x^2 = 9,$
(iv) $x^2 = 10,$　　　　(v) $x^3 = {}^-8,$　　　(vi) $x^2 = {}^-8.$

7. Write down the first ten terms of the sequence $1, 3, 5 \ldots$. Show how to select pairs of numbers from these ten terms whose sum is 20. Deduce the sum of the first ten terms of the sequence. Use a similar approach to find

    (a) the sum of the first twenty terms of this sequence,

    (b) the sum of the first fifteen terms of this sequence.

# Chapter 3
# Implication and Indices

*Implication*
If $p$ and $q$ are statements, $p \Rightarrow q$ ($q \Leftarrow p$) means that $q$ must be true if $p$ is true. $\Rightarrow$ is called the implication sign, e.g.

    (i) it is raining $\Rightarrow$ there are clouds in the sky
or     there are clouds in the sky $\Leftarrow$ it is raining.

    Note that the presence of clouds in the sky does not mean that it must be raining, so

            there are clouds in the sky $\Rightarrow$ it is raining

is incorrect.
    (ii) $x = 3 \Rightarrow x^2 = 9$.
    (iii) $x^2 = 9 \Rightarrow x = 3$ or $x = {}^-3$.

In case (iii), it is correct to reverse the implication sign since if $x = 3$ or $^-3$, $x^2$ must be equal to 9. Thus we can write

$$x = 3 \text{ or } {}^-3 \Rightarrow x^2 = 9,$$
and $$x = 3 \text{ or } {}^-3 \Leftarrow x^2 = 9.$$

These two statements can be written as the single statement

$$x = 3 \text{ or } {}^-3 \Leftrightarrow x^2 = 9.$$

*Indices*
If $x$ is a positive integer $a^x$ is defined as $x$ $a$s multiplied together, e.g.

$$3^5 = 3 \times 3 \times 3 \times 3 \times 3 = 243.$$

$a^{-x}$ is defined by $a^{-x} = 1/a^x$, e.g.

$$3^{-5} = \frac{1}{3^5} = \frac{1}{243}.$$

The function $x \rightarrow a^x$ is called an exponential function. $a$ is the exponent and $x$ the index.

10

With the above definitions the basic laws of indices are:

$$a^x \times a^y = a^{x+y}, \qquad a^x \div a^y = a^{x-y},$$

e.g $\qquad 2^7 \times 2^3 = 2^{10}, \qquad 2^3 \div 2^7 = 2^{-4}.$

If these laws are assumed to hold when $x$ is not an integer, meanings can be deduced for other expressions, e.g.

(i) $2^{\frac{1}{2}} \times 2^{\frac{1}{2}} = 2^{\frac{1}{2}+\frac{1}{2}} = 2^1 = 2 \Rightarrow 2^{\frac{1}{2}} = \sqrt{2},$

(ii) $a^0 = a^{x-x} = a^x \div a^x = 1.$

## EXERCISES

1. Insert where possible the correct sign, $\Rightarrow$, $\Leftarrow$ or $\Leftrightarrow$, in each of the following.

(a) Tom lives in Edinburgh    Tom lives in Scotland.
(b) I am British    I am Welsh.
(c) I am good at maths    I enjoy maths.
(d) 12 is an even number    12 is divisible by 2.
(e) Dogs are quadrupeds    Dogs have four legs.
(f) $x+1 = 3$    $x = 2.$
(g) $x$ is negative    $x < -2.$
(h) $P \cap Q = \varnothing$    $P \cup Q = \xi.$
(i) $ABCD$ is a rectangle    $ABCD$ is a parallelogram.
(j) 3 is a factor of 51    51 is a prime number.

2. Evaluate

(a) $2^{10}$,   (b) $5^3$,   (c) $3^{-3}$,   (d) $(^-1)^{99}$,   (e) $(^-1)^{100}$.

3. Where possible, simplify so as to leave your answer in index form.

(a) $3^2 \times 3^5$,   (b) $2^5 \times 2^{-4}$,   (c) $5^{-7} \times 5^{-2}$,
(d) $7^6 \div 7^2$,   (e) $8^4 \div 8^{-2}$,   (f) $3^2 + 3^4$,
(g) $2^6 \times 2^2/2^3$,   (h) $10^7/(10^3 \times 10^6)$,   (i) $1/5^{-2}$.

4. Calculate, justifying your answers in terms of the laws of indices,   (a) $8^{\frac{1}{3}}$,   (b) $4^{\frac{3}{2}}$,   (c) $9^{-\frac{1}{2}}$.

11

# Chapter 4
# Matrices

A matrix is a rectangular array of numbers. A matrix with $p$ rows and $q$ columns has order $p \times q$, and is said to be square if $p = q$. The matrix below has order $3 \times 4$.

$$\begin{pmatrix} 2 & 3 & -1 & \frac{1}{4} \\ 0 & -7 & 3 & 2 \\ 1 & 1 & 2 & -7 \end{pmatrix}.$$

Two matrices can be added only if they have the same order, and the addition is performed by adding corresponding elements.

$$\begin{pmatrix} 1 & 2 & 3 \\ 4 & 1 & -2 \end{pmatrix} + \begin{pmatrix} 0 & 3 & -2 \\ -5 & 0 & 4 \end{pmatrix} = \begin{pmatrix} 1 & 5 & 1 \\ -1 & 1 & 2 \end{pmatrix}.$$

A matrix is multiplied by a number when each of its elements is multiplied by that number, e.g.

$$5 \begin{pmatrix} 1 & 2 \\ 0 & -2 \\ -1 & 3 \end{pmatrix} = \begin{pmatrix} 5 & 10 \\ 0 & -10 \\ -5 & 15 \end{pmatrix}.$$

Two matrices can be multiplied together only if the number of elements in a row of the first matrix (i.e. the number of columns in the matrix) is equal to the number of elements in a column of the second matrix (i.e. the number of rows in the matrix). The answer has the same number of rows as the first matrix and the same number of columns as the second matrix. Thus if $A$ is a matrix with order $p \times q$ and $B$ is a matrix with order $r \times s$, the product $AB$ exists only if $q = r$. In this case $AB$ has order $p \times s$. If $AB$ does not exist the matrices are said to be incompatible for multiplication.

To calculate the product $AB$ we obtain the element in the

*r*th row and *s*th column of the answer by multiplying each element in the *r*th row of **A** by the corresponding element in the *s*th column of **B**, and adding the resulting products together, e.g.

$$
3\text{rd row}\begin{pmatrix} 1 & 2 & 3 & 5 \\ 2 & 1 & 4 & 3 \\ 4 & 5 & 2 & 1 \end{pmatrix}\begin{pmatrix} 2 & 3 \\ 5 & 1 \\ 2 & 4 \\ 6 & 3 \end{pmatrix}=\begin{pmatrix} & \\ & \\ & \end{pmatrix}3\text{rd row}
$$

$$
\begin{array}{ccc} & \text{2nd} & \text{2nd} \\ & \text{col.} & \text{col.} \\ 3\times4 & 4\times2 & 3\times2 \end{array}
$$

The matrices are compatible for multiplication and the answer has order $3 \times 2$. The element in the third row and second column of the answer is

$$(4 \times 3)+(5 \times 1)+(2 \times 4)+(1 \times 3) = 28.$$

The other elements are calculated in the same way, giving the answer

$$\begin{pmatrix} 48 & 32 \\ 35 & 32 \\ 43 & 28 \end{pmatrix}.$$

A zero matrix has all its elements equal to zero. If **A** is any matrix and **O** is a zero matrix,

$$\mathbf{A}+\mathbf{O} = \mathbf{O}+\mathbf{A} = \mathbf{A}, \qquad \mathbf{AO} = \mathbf{OA} = \mathbf{O},$$

where these calculations are possible.

The leading diagonal of a square matrix is the diagonal starting with the element in the first row and first column. An identity matrix is a square matrix with all elements not on the leading diagonal equal to zero and all elements on the leading diagonal equal to one.

If **A** is any matrix and **I** is an identity matrix then

$$\mathbf{AI} = \mathbf{IA} = \mathbf{A},$$

assuming multiplication is possible.

13

The transpose $\mathbf{A}'$ of a matrix $\mathbf{A}$ is the matrix obtained by interchanging the rows and columns of $\mathbf{A}$, e.g. if

$$\mathbf{A} = \begin{pmatrix} 2 & 3 & 1 & 2 \\ 1 & -1 & 2 & 0 \\ -3 & -2 & 1 & 4 \end{pmatrix} \quad \text{then} \quad \mathbf{A}' = \begin{pmatrix} 2 & 1 & -3 \\ 3 & -1 & -2 \\ 1 & 2 & 1 \\ 2 & 0 & 4 \end{pmatrix}.$$

If $\mathbf{A}$ and $\mathbf{B}$ are matrices such that $\mathbf{AB} = \mathbf{BA} = \mathbf{I}$, $\mathbf{B}$ is said to be the multiplicative inverse of $\mathbf{A}$ and is denoted by $\mathbf{A}^{-1}$. Note that the multiplications $\mathbf{AB}$ and $\mathbf{BA}$ are both possible only if $\mathbf{A}$ and $\mathbf{B}$ are both square and of the same order. If

$$\mathbf{A} = \begin{pmatrix} a & b \\ c & d \end{pmatrix} \quad \text{then} \quad \mathbf{A}^{-1} = \begin{pmatrix} d/\Delta & -b/\Delta \\ -c/\Delta & a/\Delta \end{pmatrix}$$

where $\Delta = ad - bc$, $\Delta$ is called the determinant of the matrix. For example, if

$$\mathbf{A} = \begin{pmatrix} 4 & 1 \\ 3 & 2 \end{pmatrix}$$

then

$$\Delta = (4 \times 2) - (1 \times 3) = 5$$

and

$$\mathbf{A}^{-1} = \begin{pmatrix} 2/5 & -1/5 \\ -3/5 & 4/5 \end{pmatrix}.$$

If $\Delta = 0$ the matrix has no inverse and is called singular.

## EXERCISES

1. A treasure hunt is organized by hiding a number of red, yellow, blue and black discs. Three teams search for these. Team $A$ finds two red, five yellow, twelve blue and four black discs. Team $B$ finds one red, eight yellow, nine blue and two black discs. Team $C$ finds no red, thirteen yellow, three blue and five black discs. Express this information as a $3 \times 4$ matrix.

The teams are then informed that each red disc is worth ten

points, each yellow disc four points, each blue disc one point and each black disc carries a penalty point. By performing a suitable matrix multiplication find the total score of each team.

2. A housewife buys six apples, three oranges and three pears on Monday, four oranges on Tuesday, three apples and four pears on Wednesday, two apples on Thursday and four oranges, six apples and two pears on Friday. Represent this information as a $5 \times 3$ matrix.

If each apple costs 2p, each pear costs $2\frac{1}{2}$p and each orange cost 3p write down a matrix product which will give the total amount that she spends each day on fruit. Find her total expenditure on apples, oranges and pears for the five days.

3. Find the answer to the only multiplication possible using any two of the three matrices $\mathbf{A}$, $\mathbf{B}$ and $\mathbf{C}$ where

$$\mathbf{A} = \begin{pmatrix} 3 & 2 \\ 1 & 0 \end{pmatrix}, \qquad \mathbf{B} = \begin{pmatrix} 3 & 2 \\ 1 & 0 \\ 4 & 1 \end{pmatrix}, \qquad \mathbf{C} = \begin{pmatrix} 3 \\ 1 \\ 1 \end{pmatrix}.$$

4. The matrices $\mathbf{A}$, $\mathbf{B}$ and $\mathbf{C}$ are given by

$$\mathbf{A} = \begin{pmatrix} -1 & 2 \\ 3 & -2 \end{pmatrix}, \qquad \mathbf{B} = \begin{pmatrix} 3 & -4 \\ 1 & 1 \end{pmatrix}, \qquad \mathbf{C} = \begin{pmatrix} 2 & -1 \\ -3 & 4 \\ -5 & 0 \end{pmatrix}.$$

Calculate where possible

(a) $3\mathbf{C}$,
(b) $\mathbf{A} + \mathbf{C}$,
(c) $\mathbf{B} + \mathbf{A}$,
(d) $\mathbf{AB}$,
(e) $\mathbf{BC}$,
(f) $\mathbf{CA}$,
(g) $2\mathbf{A} + 3\mathbf{B}$,
(h) $\mathbf{C}(\mathbf{A} + \mathbf{B})$,
(i) $\mathbf{C}'$,
(j) $\mathbf{A}^{-1}$,
(k) $\mathbf{B}^{-1}$,
(l) $\mathbf{CC}'$.

5. Find the inverses, where possible, of the following matrices.

(a) $\begin{pmatrix} 8 & -3 \\ -5 & 2 \end{pmatrix}$,
(b) $\begin{pmatrix} 2 & 3 \\ -2 & 1 \end{pmatrix}$,
(c) $\begin{pmatrix} 3 & 1 \\ 2 & -3 \end{pmatrix}$,

(d) $\begin{pmatrix} -2 & -4 \\ -5 & -10 \end{pmatrix}$,
(e) $\begin{pmatrix} 0 & 3 \\ 2 & 1 \end{pmatrix}$,
(f) $\begin{pmatrix} a & b \\ \dfrac{2}{b} & \dfrac{3}{a} \end{pmatrix}$.

15

Where an inverse exists, check your answer by matrix multiplication.

6. The matrices **P**, **Q** and **R** are given by

$$\mathbf{P} = \begin{pmatrix} 2 & -1 \\ 3 & -2 \end{pmatrix}, \qquad \mathbf{Q} = \begin{pmatrix} -1 & 5 \\ 2 & -3 \end{pmatrix}, \qquad \mathbf{R} = \begin{pmatrix} 0 & 3 \\ -2 & 1 \end{pmatrix}.$$

Calculate

 (a) **P**+**Q** and **Q**+**P**,     (b) **PQ** and **QP**,
 (c) **P**(**Q**+**R**) and **PQ**+**PR**,  (d) (**PQ**)**R** and **P**(**QR**).

State the property of matrices illustrated by your answer to each part of this question.

7. If **A** is any matrix and **A′** is its transpose, explain why the multiplication **AA′** is always possible. What can you say about the order of the matrix **AA′**?

8. A message is coded by replacing each letter of the alphabet by the number corresponding to its position in the alphabet ($a = 1, b = 2$, etc.). These numbers are taken in pairs, expressed as column vectors, multiplied by the matrix

$$\begin{pmatrix} 2 & 1 \\ -1 & 1 \end{pmatrix}$$

and put back in sequence. The message 21   −3   40   4 is found. Decode this message.

9. The matrices **A**, **B** and **C** are given by

$$\mathbf{A} = \begin{pmatrix} 2 & 3 \\ 1 & 1 \end{pmatrix}, \qquad \mathbf{B} = \begin{pmatrix} 3 & 2 \\ 4 & 3 \end{pmatrix}, \qquad \mathbf{C} = \begin{pmatrix} 2 & -3 \\ 1 & -2 \end{pmatrix}.$$

Solve the following equations, where possible.

 (a) **A**+**X** = **C**,   (b) **AX** = **B**,    (c) **XA** = **B**,
 (d) **AX**+**B** = **C**,  (e) **A**(**X**+**B**) = **C**,  (f) **AX**−**B** = **CX**.

10. If the matrix $\begin{pmatrix} a & b \\ c & d \end{pmatrix}$ is singular, evaluate the matrix

product $\begin{pmatrix} a & b \\ c & d \end{pmatrix} \begin{pmatrix} d & -b \\ -c & a \end{pmatrix}$. Hence find, where possible, solutions other than the zero matrix of the following equations:

(a) $\begin{pmatrix} 1 & 2 \\ 2 & 4 \end{pmatrix} \mathbf{X} = \mathbf{O}$,   (b) $\begin{pmatrix} -2 & 6 \\ -1 & 3 \end{pmatrix} \mathbf{X} = \mathbf{O}$,

(c) $\begin{pmatrix} 3 & -2 \\ 6 & 4 \end{pmatrix} \mathbf{X} = \mathbf{O}$.

In general when has the equation $\mathbf{AX} = \mathbf{O}$ a non-zero solution? Is there more than one such solution? If $\mathbf{A}, \mathbf{B}$ and $\mathbf{X}$ are matrices is it true that

$$(\mathbf{X} - \mathbf{A})(\mathbf{X} - \mathbf{B}) = \mathbf{O} \Rightarrow \mathbf{X} = \mathbf{A} \text{ or } \mathbf{X} = \mathbf{B}?$$

# Chapter 5
# Relations and Functions

In Fig. 5.1 the arrows represent the relation, or mapping, 'spent his holiday at'. Thus Arthur spent his holiday at Skegness, Francis spent holidays at Hastings and Clacton, etc. No one visited London. In this example:

  (i) $A$ is the domain of the mapping,
 (ii) Hastings is the image of Dick, {Hastings, Clacton} is the image set of Francis,
(iii) the range is {Blackpool, Skegness, Hastings, Margate, Clacton}, i.e. the set of images of elements of the domain,
 (iv) the inverse relation would map $B$ onto $A$, and would be represented by Fig. 5.1 with the arrows reversed in direction. This relation can be described 'was stayed at by'.

A relation is a function if each element of the domain is mapped onto one and only one element of the range. The relation in Fig. 5.1 would have been a function if Francis had visited only one of Clacton or Hastings, but not both.

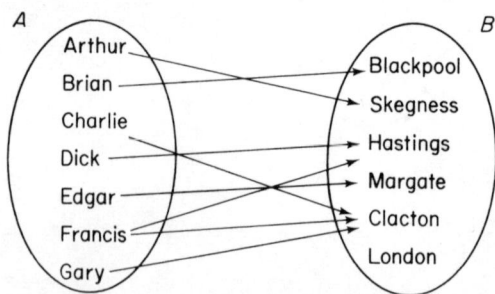

FIG. 5.1

If $f$ is a function mapping a set $X$ to a set $Y$, written $f: X \to Y$, and $y$ is the image of $x$ under $f$, we write

$$f(x) = y, \quad f: x \to y \quad \text{or} \quad x \xrightarrow{f} y.$$

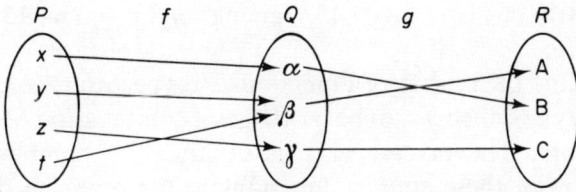

FIG. 5.2

Fig. 5.2 represents the functions $f:P \to Q$ and $g:Q \to R$. Here the arrows can be traced across the diagram to give the sequence

$$x \xrightarrow{f} \alpha \xrightarrow{g} B.$$

This can be written $g\{f(x)\} = g(\alpha) = B$, or $gf(x) = B$. The single function $gf$ is shown in Fig. 5.3.

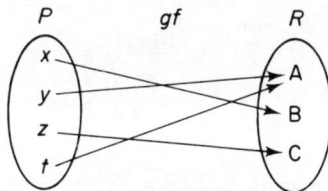

FIG. 5.3

Any function has an inverse relation, which is not necessarily itself a function, e.g. in Fig. 5.2 the inverse relation of $f$ is not a function as it maps $\beta$ onto both $y$ and $t$, whilst the inverse relation of $g$ is a function, in this case written $g^{-1}:R \to Q$. The function $g^{-1}g$ is the identity function and maps each element of $Q$ onto itself.

Functions with domain and range whose elements are numbers are often written in the form $x \to f(x)$, e.g.

$$f:x \to x+7, \qquad g:x \to 2x-1.$$

Here $g(3) = 5$, $fg(3) = f(5) = 12$. Since $g(3) = 5$, therefore $g^{-1}(5) = 3$. We have

$$fg(x) = f(2x-1) = (2x-1)+7 = 2x+6,$$

giving

$$fg:x \to 2x+6.$$

19

Similarly $gf(x) = 2x+13$, giving $gf: x \to 2x+13$. Thus $gf \neq fg$.

To find the inverse of a function $f$ it is necessary to consider the way in which $f$ can be written as a combination of simpler functions. The inverse of $f$ is obtained by combining the inverses of these simpler functions in the opposite order to that in which the original functions combined to give $f$.

For example, the function $f: x \to 3x+7$ can be considered as the functions 'multiply by 3' followed by 'add 7'. Note that the order here is important, 'add 7' followed by 'multiply by 3' would give the function $x \to 3(x+7)$. The inverses of these two simple functions are easily seen to be 'divide by 3' and 'subtract 7'. Thus the inverse $f^{-1}$ is the function 'subtract 7' followed by the function 'divide by 3'. This can be written

$$f: x \xrightarrow{\substack{multiply \\ by\ 3}} 3x \xrightarrow{add\ 7} 3x+7,$$

$$f^{-1}: x \xrightarrow{\substack{subtract \\ 7}} x-7 \xrightarrow{\substack{divide \\ by\ 3}} \frac{x-7}{3}.$$

Note that the functions 'subtract from' and 'divide into' are self-inverse.


## EXAMPLES

1. State which of the following relations are functions, and which have inverse functions. (In each case take the domain to be the set of real numbers.)

(a) $x \to 2$,  (b) $x \to 2-3x$,
(c) $x \to x^2$,  (d) $x \to \sqrt{(1-x^2)}$,
(e) $x \to \sin x$,  (f) $x \to$ the angle whose cosine is $x$.

2. Given $f: x \to 2x-1$ and $g: x \to 5-3x$, find

(a) $f(2)$,  (b) $f(-1)$,  (c) $g(2)$,  (d) $g(-2)$,
(e) $fg(2)$,  (f) $gf(2)$,  (g) $ff(-1)$,  (h) $gg(-2)$,
(i) $fgf(2)$,  (j) $gfg(0)$.

3. State the range of each of the following functions for the domain given:
  (a) $x \to \frac{1}{2}x - 1$, domain $= \{0, 1, 2, 3, 4\}$
  (b) $x \to x^2$, domain $= \{-2, -1, 0, 1, 2\}$
  (c) $x \to 0$, domain $= \{$real numbers$\}$
  (d) $x \to \frac{1}{x}$, domain $= \{x : x \geqslant 1\}$
  (e) $x \to \sin x°$, domain $= \{x : 0 \leqslant x \leqslant 360\}$
  (f) $x \to x^2 + 6$, domain $= \{$real numbers$\}$

4. State a domain for which each of the following relations is a function:
  (a) $x \to \frac{1}{3}(x + 2)^3$, (b) $x \to \sqrt{(9 - x^2)}$, (c) $x \to 2^x$
  (d) $x \to \sqrt{(x^2 - 16)}$, (e) $x \to x/(x+1)$, (f) $x \to \log_{10} x$,
  (g) $x \to \{$prime factors of $x\}$.

5. Find the inverse of each of the following functions:
  (a) $x \to 3x + 4$, (b) $x \to (x-3)/4$,
  (c) $x \to \frac{1}{2}x - 1\frac{1}{2}$, (d) $x \to \frac{1}{3}(2x - 1)$,
  (e) $x \to 1 - 4x$, (f) $x \to (1/x) + 2$,
  (g) $x \to 6/(4-x)$, (h) $x \to (x+2)^2$.

6. Given $f : x \to x - 2$, $g : x \to \frac{1}{3}x$ and $h : x \to 2(x+2)$ express each of the following in the form $x \to \ldots$

(a) $f^{-1}$, (b) $g^{-1}$, (c) $h^{-1}$, (d) $fg$, (e) $gf$,
(f) $fh$, (g) $hf$, (h) $gh$, (i) $hg$, (j) $fg^{-1}$,
(k) $(fg)^{-1}$, (l) $g^{-1}f^{-1}$, (m) $ff^{-1}$, (n) $fgf^{-1}$, (o) $fgh$.

7. The relation 'was born in', which maps $\{$Angus, Brenda, Cynthia, Dick, Eric$\}$ onto $\{$Glasgow, London, Penzance, Nottingham$\}$, is represented by the matrix

|  | Gl. | Lo. | Pe. | No. |
|---|---|---|---|---|
| Angus | 1 | 0 | 0 | 0 |
| Brenda | 0 | 1 | 0 | 0 |
| Cynthia | 0 | 0 | 1 | 0 |
| Dick | 0 | 1 | 0 | 0 |
| Eric | 0 | 0 | 0 | 1 |

21

In this matrix the 1 in the fourth row and second column denotes that Dick was born in London, etc. Represent this relation diagrammatically. Is the relation a function? How could this be recognized by examining the matrix? Is it possible to recognize from the matrix whether the inverse relation is a function? What is the connection between the matrix representing the relation and the matrix representing the inverse relation?

8. $x$ and $y$ are natural numbers and $f$ is the relation mapping the ordered pair $(x, y)$ onto the highest common factor of $x$ and $y$.

(a) Is $f$ a function?
(b) Find the image of $(15, 20)$ and $(72, 180)$ and $f$.
(c) If $(x, 42) \xrightarrow{f} 14$, state four possible values of $x$.
(d) Is the inverse relation a function?
(e) What can you say about $x$ and $y$ if $(x, y) \xrightarrow{f} x$?

9. The function $d$ maps a natural number onto the number of factors that number possesses, e.g. 14 has factors $1, 2, 7, 14$ so $d(14) = 4$.

(a) Find $d(13)$, $d(24)$, $d(30)$ and $d(100)$.
(b) If $d(x) = 2$, what can you say about $x$?
(c) Is there a number for which $d(x) = x$?
(d) Find $d(4)$, $d(8)$, $d(16)$, $d(2^n)$.

# Chapter 6
# Translations and Vectors

When a translation is applied to an object all points of the object move the same distance in the same direction.

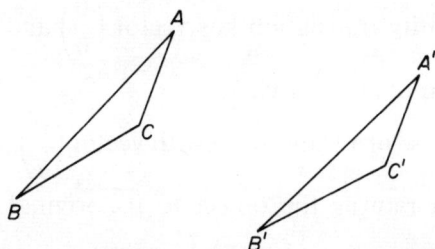

FIG. 6.1

In Fig. 6.1 $\triangle A'B'C'$ is the image of $\triangle ABC$ under a translation. By definition the lines $AA'$, $BB'$ and $CC'$ must be equal in length and parallel. Thus the length and direction of $AA'$ completely define the translation. We can say that the vector $\mathbf{AA'}$ defines the translation. Hence $\mathbf{AA'} = \mathbf{BB'} = \mathbf{CC'}$, since each represents the same translation.

It follows from the definition that if two translations are applied in succession they will be equivalent to a third translation. In Fig. 6.2,

$$\mathbf{AA'} + \mathbf{A'A^*} = \mathbf{AA^*}$$
$$(\text{or} \quad \mathbf{BB'} + \mathbf{CC^*} = \mathbf{AA^*} \quad \text{etc.}).$$

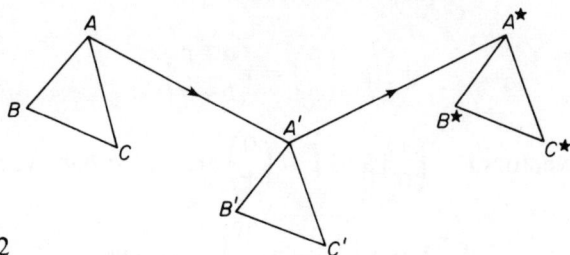

FIG. 6.2

23

When using cartesian coordinates, the column vector $\begin{pmatrix} a \\ b \end{pmatrix}$ describes the translation resulting from a translation through $a$ parallel to the line $y = 0$ (the $x$-axis) followed by a translation through $b$ units parallel to the line $x = 0$ (the $y$-axis).

$$\begin{pmatrix} a \\ b \end{pmatrix} + \begin{pmatrix} c \\ d \end{pmatrix} = \begin{pmatrix} a+c \\ b+d \end{pmatrix} \qquad n\begin{pmatrix} a \\ b \end{pmatrix} = \begin{pmatrix} na \\ nb \end{pmatrix}.$$

The identity translation has vector $\begin{pmatrix} 0 \\ 0 \end{pmatrix}$ and does not alter the position of any object.

The inverse of a translation with vector $\begin{pmatrix} p \\ q \end{pmatrix}$ (i.e. the translation which returns the object to its original position) is a translation with vector $\begin{pmatrix} -p \\ -q \end{pmatrix}$,

$$\begin{pmatrix} p \\ q \end{pmatrix} + \begin{pmatrix} -p \\ -q \end{pmatrix} = \begin{pmatrix} 0 \\ 0 \end{pmatrix}.$$

The position vector of a point is the vector of the translation which would map the origin onto the point. The position vector of the point $(p, q)$ is $\begin{pmatrix} p \\ q \end{pmatrix}$.

The translation with vector $\begin{pmatrix} p \\ q \end{pmatrix}$ maps the point whose position vector is $\begin{pmatrix} a \\ b \end{pmatrix}$ onto the point whose position vector is

$$\begin{pmatrix} a \\ b \end{pmatrix} + \begin{pmatrix} p \\ q \end{pmatrix} = \begin{pmatrix} a+p \\ b+q \end{pmatrix}.$$

The vectors $\mathbf{i} = \begin{pmatrix} 1 \\ 0 \end{pmatrix}$ and $\mathbf{j} = \begin{pmatrix} 0 \\ 1 \end{pmatrix}$ are called base vectors.

$$\begin{pmatrix} x \\ y \end{pmatrix} = x\begin{pmatrix} 1 \\ 0 \end{pmatrix} + y\begin{pmatrix} 0 \\ 1 \end{pmatrix} = x\mathbf{i} + y\mathbf{j}.$$

## EXERCISES

1. The translation with column vector $\begin{pmatrix} 1 \\ -2 \end{pmatrix}$ is applied to the square whose vertices are $A(2, 1)$, $B(1, -2)$, $C(-2, -1)$ and $D(-1, 2)$. Give the coordinates of the vertices of the image $A'B'C'D'$.

2. State the column vector of the translation which maps the first point of each of the following pairs onto the second point.

    (a) $(2, 1)$   $(5, 3)$,            (b) $(-3, 0)$   $(-1, 6)$,
    (c) $(-3, 4)$   $(-1, 6)$,       (d) $(5, 3)$   $(2, 1)$.

What is the relationship between the transformations in (a) and (d)?

3. The quadrilateral with vertices $A'(1, 2)$, $B'(4, 3)$, $C'(3, 0)$ and $D'(2, -1)$ is the image of the quadrilateral $ABCD$ under the translation with column vector $\begin{pmatrix} 3 \\ 1 \end{pmatrix}$. Give the coordinates of the points $A, B, C$ and $D$.

4. A triangle has vertices $A(1, 1)$, $B(-1, 2)$ and $C(2, 3)$. Find the coordinates of the vertices of the image $A'B'C'$ of the triangle under a translation which maps the point $(1, 2)$ onto the point $(-1, 0)$.

    Give the transformation which maps the image of $\triangle ABC$ under the translation with vector $\begin{pmatrix} 3 \\ -2 \end{pmatrix}$ onto $\triangle A'B'C'$.

5. If $\mathbf{a} = \begin{pmatrix} 2 \\ -3 \end{pmatrix}$ and $\mathbf{b} = \begin{pmatrix} -1 \\ 4 \end{pmatrix}$, simplify

    (a) $\mathbf{a} + \mathbf{b}$,         (b) $\mathbf{a} - \mathbf{b}$,         (c) $-\mathbf{a}$,
    (d) $2\mathbf{a}$,            (e) $3\mathbf{b}$,            (f) $3\mathbf{a} + 2\mathbf{b}$,
    (g) $3\mathbf{b} - 2\mathbf{a}$,     (h) $2\mathbf{a} - 4\mathbf{b}$.

6. A ship leaving port moves 10 km on a bearing 032°, and

then changes course to move a further 15 km on a bearing 076°. Use a scale drawing to find

  (a) its distance from the port after the second stage of the journey,
  (b) its bearing from the port at this time.

7. In Fig. 6.3 $C$, $D$ and $E$ are the mid-points of the sides of the triangle $ABO$ and $OECD$ is a parallelogram. If $\mathbf{OE} = \mathbf{e}$ and $\mathbf{OD} = \mathbf{d}$ express the following vectors in terms of $\mathbf{e}$ and $\mathbf{d}$.

  (a) **EA**,   (b) **EC**,   (c) **OC**,   (d) **OB**,
  (e) **DE**,   (f) **AD**,   (g) **AB**,   (h) **BE**.

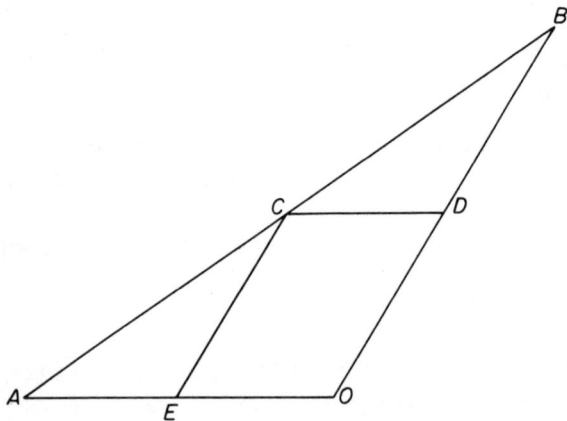

FIG. 6.3

8. A man takes a trip in a balloon and after an hour he has drifted 10 km to the north and 5 km to the east in attaining a height of 2 km. His position at this time is represented by the vector

$$\begin{pmatrix} 10 \\ 5 \\ 2 \end{pmatrix}.$$

In the next half-hour he drifts a further 2 km north and 3 km west, and rises another $\frac{1}{2}$ km. Give the vector which

  (a) describes the second stage of his journey,
  (b) represents his overall position.

If the balloon returns to the ground 6 km due north of its starting-point after $2\frac{1}{2}$ hours in the air, give a vector to describe the last hour of the journey.

9. Fig. 6.4 represents a cube.

  (i) Name the vectors equivalent to  (a) **AB**,    (b) **AE**.
 (ii) Simplify
      (a) **AD** + **DE**,           (b) **AB** + **AE**,    (c) **AH** + **ED**,
      (d) **DE** + **AD** + **EH**,    (e) **AB** + **CD**.
(iii) State the position of $X$ if **AD** + **AH** = 2**AX**.

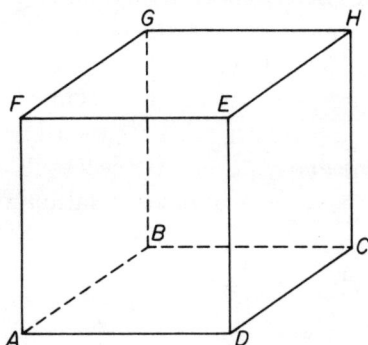

FIG. 6.4

27

# Chapter 7
# Reflection and Rotation

In Fig. 7.1, if $\triangle A'B'C'$ is the image of $\triangle ABC$ after reflection in the line $l$, the line joining any point on the object to the corresponding point on the image is bisected by the line $l$ and is perpendicular to $l$ ($l$ is the mediator of $AA'$, $BB'$, ...). A reflection is completely specified by naming the mirror line.

Any point lying on $l$ has its position unchanged under reflection in $l$. If $X$, $Y$ have images $X'$, $Y'$ after reflection in $l$, the lines $XY$ and $X'Y'$ are equally inclined to $l$ and meet on $l$ if produced. Any reflection is its own inverse as reflecting twice in the same line restores an object to its original position.

If $A'$ is the image of $A$ under a rotation about a centre $O$,

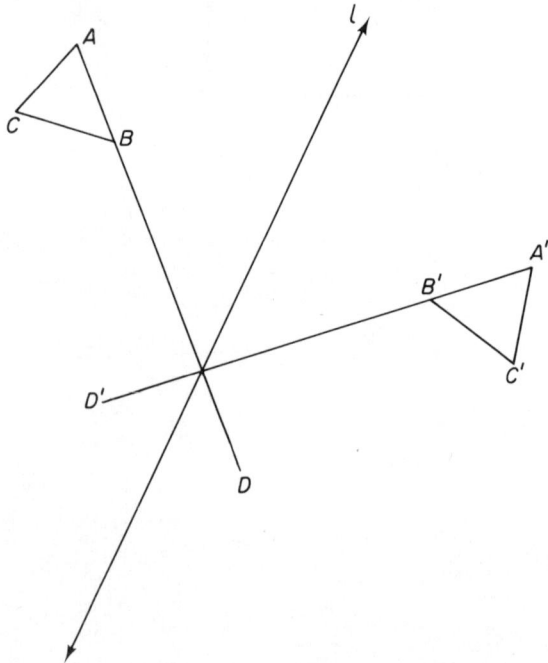

Fig. 7.1

28

$OA' = OA$ (see Fig. 7.2). If $B'$ is the image of $B$ under the same rotation, $A\hat{O}A' = B\hat{O}B'$. $O$ is the centre of rotation and if the angle of rotation is $\theta$, $B\hat{O}B' = C\hat{O}C' = A\hat{O}A' = \theta$.

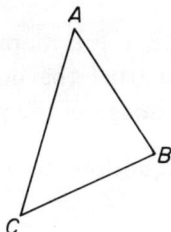

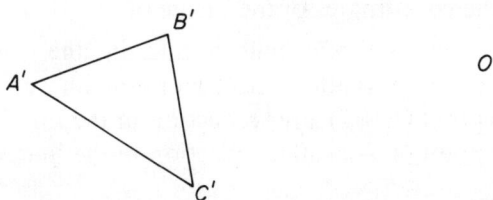

FIG. 7.2

A rotation is completely specified by giving the centre and angle of rotation. The angle of rotation is positive when measured in the anticlockwise direction. Any line of the object is turned through the same angle (i.e. through the angle of rotation).

The identity rotation is a rotation through $0°$ (or any multiple of $360°$) about any centre. The inverse of a rotation centre $C$, angle $\theta$ is a rotation centre $C$, angle $-\theta$.

## EXERCISES

1. A triangle has vertices $A(1, 1)$, $B(3, 1)$, $C(3\ 2)$. Find the co-ordinates of the vertices of the image of the triangle under each of the following transformations:

29

(a) Rotation through $90°$ about $(0, 0)$,
(b) Rotation through $180°$ about $(0, 1)$,
(c) Reflection in $y = 0$,
(d) Reflection in $x = 2$.

2. The point $(2, 1\frac{1}{2})$ is mapped onto the point $(-1, 3)$ by a reflection. Construct the mirror line for this reflection and hence find the image of the point $(-1, 5)$ under the reflection.

3. Give the equation of the mirror line for the reflection which maps

(a) the point $(1, 1)$ onto the point $(1, 5)$,
(b) the point $(-5, 2)$ onto the point $(1, 6)$,
(c) the point $(1, 0)$ onto the point $(-1, 3)$,
(d) the point $(-2, 2)$ onto the point $(6, -2)$.

4. Give the coordinates of the image of

(a) the point $(3, 1)$ after reflection in the line $y = x + 1$,
(b) the point $(5, 4)$ after reflection in the line $y + 2x = 4$,
(c) the point $(1, -2)$ after reflection in the line $2y = 3x - 7$,
(d) the point $(4, -1)$ after reflection in the line $3y = x + 3$.

5. Under a rotation the point $(-5, 5)$ is mapped onto the point $(1, 5)$. Show on a diagram the line on which the centre of rotation must lie.

The same rotation maps the point $(-4, 8)$ onto the point $(2, 2)$. Find, by drawing, the centre of rotation and the angle of rotation.

6. State the angle and centre of the rotation which maps

(a) the points $(-4, 4)$, $(-1, 5)$ onto the points $(3, 1)$, $(2, -2)$ respectively,
(b) the points $(-1, 7)$, $(-5, 2)$ onto the points $(3, -5)$, $(7, 0)$ respectively,
(c) the points $(3, 21)$, $(-7, 17)$ onto the points $(-9, 15)$, $(-13, 5)$ respectively.

7. Fig. 7.3 shows a regular octagon. Describe a single trans-

formation which will effect the following mappings. ($XY \rightarrow$ $PQ$ indicates that $X \rightarrow P$ and $Y \rightarrow Q$.)

(a) $AB \rightarrow BC$,        (b) $GH \rightarrow CD$,

(c) $GH \rightarrow DC$,        (d) $AG \rightarrow BD$,

(e) $BF \rightarrow CG$,        (f) $ABCDEFGH \rightarrow HGFEDCBA$.

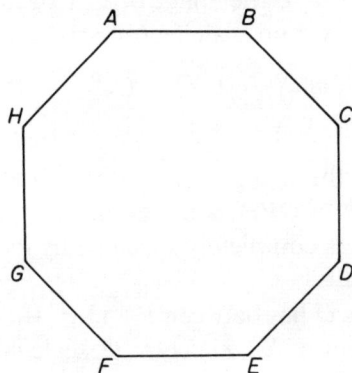

FIG. 7.3

8. The triangle whose vertices are $A(1, 3)$, $B(2, 3)$, $C(2, 5)$ is reflected in the line $y = x$. Find the coordinates of the vertices of the image $A'B'C'$.

The mirror line is now rotated through $90°$ about the point $(0, 1)$ onto the line $l'$. Give the equation of $l'$.

Find the coordinates of the image $A*B*C*$ of $\triangle ABC$ after reflection in $l'$.

State the single transformation which maps $\triangle A'B'C'$ onto $\triangle A*B*C*$.

31

# Chapter 8
# Enlargement and Shearing

In Fig. 8.1 $\triangle X'Y'Z'$ is the image of $\triangle XYZ$ under an enlargement with centre $C$ and scale factor 3, i.e.

$$\frac{CX'}{CX} = \frac{CY'}{CY} = \frac{CZ'}{CZ} = 3.$$

In general $P'$ is the image of $P$ under an enlargement centre $O$, scale factor $k$, if $OPP'$ is a straight line and $OP'/OP = k$. An enlargement is completely specified by its centre and scale factor.

If $k$ is negative $O$ lies between $P$ and $P'$. (In Fig. 8.1 $X_2Y_2Z_2$ is the enlargement of $XYZ$, centre $C$, scale factor $-2$.)

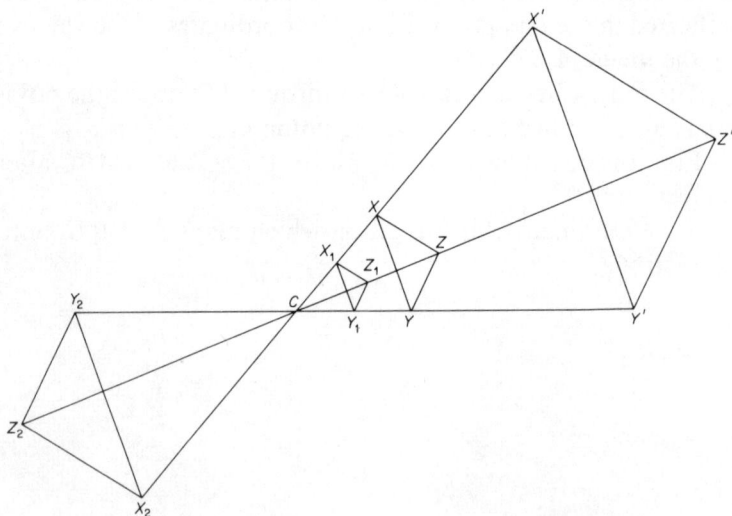

FIG. 8.1

If $-1 < k < 1$ the image is smaller than the original object. (In Fig. 8.1 $X_1 Y_1 Z_1$ is the enlargement of $XYZ$, centre $C$, scale factor $\frac{1}{2}$.)

If $P'Q'$ is the image of any line segment $PQ$ under an enlargement centre $O$, scale factor $k$, $P'Q'/PQ = k$ and $P'Q'$ is parallel to $PQ$. (In Fig. 8.1

$$\frac{X'Y'}{XY} = 3, \ X'Y'//XY; \quad \frac{X_2 Y_2}{XY} = 2, \ X_2 Y_2 //XY.)$$

An object and its image under any enlargement are similar (i.e. the same shape).

The inverse of an enlargement centre $O$, scale factor $k$ is an enlargement centre $O$, scale factor $1/k$. (In Fig. 8.1 $\triangle XYZ$ is the image of $\triangle X'Y'Z'$ under the enlargement centre $C$, scale factor $\frac{1}{3}$.)

Under an enlargement of scale factor $k$ a figure has its area increased by a factor $k^2$. A solid has its volume increased by a factor $k^3$.

In Fig. 8.2 $ABC'D'$ is the image of $ABCD$ under a shear with invariant line $AB$. The line $AB$ remains unchanged under the

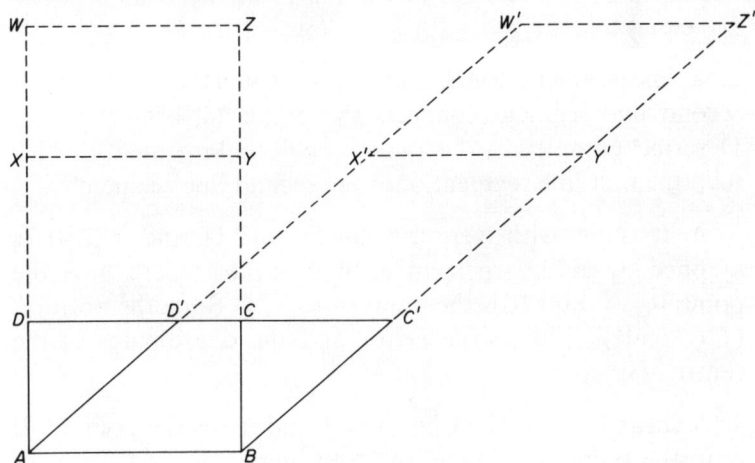

FIG. 8.2

transformation and all points move parallel to *AB* through a distance proportional to their distance from *AB*.

A shear is completely specified by giving the invariant line and the image of one point not on the invariant line (e.g. *D* is mapped onto *D'*).

Note that $W'X'Y'Z'$ is the image of $WXYZ$ under the same shear.

If $P'$ is the image of $P$ under a certain shear and $Q'$ is the image of $Q$ under the same shear, $P'Q'$ and $PQ$ intersect on the invariant line.

Shearing preserves area, i.e. the areas of a figure and its image under a shear are equal.

The inverse of a shear with invariant line $ST$ which maps a point $U$ onto a point $U'$ is the shear with invariant line $ST$ which maps the point $U'$ onto the point $U$.

## EXERCISES

1. A triangle with vertices $A(0, 1)$, $B(0, -1)$ and $C(2, 2)$ is enlarged by a factor $k$ about the centre $(-1, 0)$. Find the co-ordinates of the vertices of the image of the triangle when $k$ is equal to    (a) 2,    (b) $\frac{1}{2}$,    (c) $-2$.

2. A line segment connects the points $(0, 1)$, and $(0, -1)$. A second line segment connects the points $(4, 4)$ and $(4, -4)$. Describe completely the two possible enlargements which map the first line segment onto the second line segment.

3. A triangle with vertices $A(4, 5)$, $B(3, 1)$ and $C(2, 4)$ is mapped by an enlargement onto $\triangle A'B'C'$ where $A'$ is the point $(0, -\frac{1}{2})$ and $B'$ is the point $(-\frac{1}{2}, -2\frac{1}{2})$. State the position of $C'$ and give the scale factor and the coordinates of the centre of enlargement.

4. A shear has invariant line $x = 1$ and maps the point $(3, 2)$ onto the point $(1, 2)$. State the coordinates of the vertices of the images of each of the following figures under this shear:

34

    (a) the rectangle with vertices (0, 1), (0, 2), (2, 2) and (2, 1),

    (b) the rectangle with vertices (0, 0), (−1, 0), (−1, 2) and (0, 2).

5. Show on a diagram the invariant line for each of the following shears:

    (a) The shear which maps the points (−1, −2), (1, 2) onto the points (2, −2), (10, 2) respectively.

    (b) The shear which maps the points (1, 3), (1, −3) onto the points (2, 2), (−1, −1) respectively.

    (c) The shear which maps the points (−6, 6), (0, 5) onto the points (6, 12), (4, 7) respectively.

6. Find the area of the quadrilateral whose vertices are $A(1, 0)$, $B(1, 2)$, $C(2, 5)$ and $D(3, 4)$ by applying separate shears to the triangles $ACD$ and $ABD$. (Take $BD$ as invariant line in each case.)

7. A map of a town is drawn to a scale 1/10 000.

    (a) If the distance from the town hall to the library is 750 m, how far apart are the corresponding points on the map?

    (b) On the map I measure the distance from my house to school as 13 cm. How far is the school from my house?

    (c) The area of a park on the map is 12 cm². Find the actual area of the park.

8. An architect builds a model of a block of flats on a scale 1/50.

    (a) If the block of flats is to be 34 m high, find the height of the model.

    (b) If the model has 232 windows, how many windows will the block of flats have?

    (c) If the area of floor in a flat is to be 95 m², find the area of the floor of a flat in the model.

    (d) It is proposed to build a swimming-pool in the grounds of the block of flats. If the volume of the model swimming-pool is 2400 cm³, find the volume of water that will be required to fill the actual pool.

# Chapter 9
# Transformations and Matrices

A transformation can be considered as a function mapping the plane onto itself. Thus if **S** and **T** are transformations, the transformation **ST** is the single transformation equivalent to applying first **T** then **S**.

An isometry is a transformation which preserves shape and size. Thus the image of a figure under an isometry is congruent to the figure. Since combining any two of the set of isometries already discussed, i.e.{reflection, rotation, translation} does not always yield a member of the set, it is useful to define a fourth isometry, the glide-reflection, as a reflection followed by a translation parallel to the mirror line. {Reflection, rotation, translation, glide-reflection} is closed under the operation of combining transformations, i.e. any two elements of the set combined together give an element of the set (see the table).

In Fig. 9.1, $F_2$ and $F_3$ are directly congruent since $F_3$ can

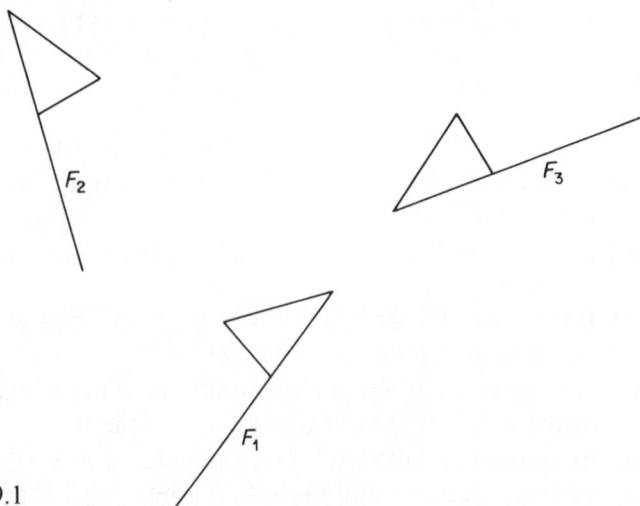

FIG. 9.1

36

|  | Translation | Reflection | Rotation | Glide-reflection |
|---|---|---|---|---|
| Translation | Translation | Glide-reflection | Rotation | Glide-reflection |
| Reflection | Glide-reflection | Translation through twice distance between mirrors (if mirrors parallel) Rotation through twice angle between mirrors, centre at intersection of mirrors (if not parallel) | Glide-reflection | Translation or rotation |
| Rotation | Rotation | Glide-reflection | Rotation, or translation if angles of rotations are equal and opposite | Glide-reflection |
| Glide-reflection | Glide-reflection | Translation or rotation | Glide-reflection | Translation or rotation |

be moved in the plane of the paper to the position of $F_2$. $F_1$ is indirectly congruent to $F_2$ since $F_1$ cannot be moved in this way to $F_2$. Thus $F_2$ can be mapped onto $F_3$ by a direct isometry, $F_2$ onto $F_1$ by an indirect isometry. It follows that rotations and translations are direct isometries, reflections and glide-reflections are indirect isometries. The following table for combining direct and indirect isometries is useful.

|          | Direct   | Indirect |
|----------|----------|----------|
| Direct   | Direct   | Indirect |
| Indirect | Indirect | Direct   |

The image of the point with position vector $\begin{pmatrix} x \\ y \end{pmatrix}$ under the transformation with matrix $\begin{pmatrix} a & b \\ c & d \end{pmatrix}$ is

$$\begin{pmatrix} a & b \\ c & d \end{pmatrix} \begin{pmatrix} x \\ y \end{pmatrix} = \begin{pmatrix} ax+by \\ cx+dy \end{pmatrix}.$$

The matrices corresponding to some common transformations are shown in the table.

If **S** and **T** are transformations with matrices **A** and **B** respectively, the matrix of the transformation **ST** is **AB**. The matrix of the inverse transformation of **S** is $\mathbf{A}^{-1}$; e.g. if **S** is the rotation about the point $(0,0)$ through $90°$, then

$$\mathbf{A} = \begin{pmatrix} 0 & -1 \\ 1 & 0 \end{pmatrix}$$

and if **T** represents reflection in the line $x = 0$, then

$$\mathbf{B} = \begin{pmatrix} -1 & 0 \\ 0 & 1 \end{pmatrix}.$$

Thus $\mathbf{AB} = \begin{pmatrix} 0 & -1 \\ 1 & 0 \end{pmatrix} \begin{pmatrix} -1 & 0 \\ 0 & 1 \end{pmatrix} = \begin{pmatrix} 0 & -1 \\ -1 & 0 \end{pmatrix}$, and

**AB** is the matrix of the transformation **ST**. Hence the single transformation equivalent to **ST** is reflection in the line $x + y = 0$.

| Transformation | Matrix |
|---|---|
| Identity | $\begin{pmatrix} 1 & 0 \\ 0 & 1 \end{pmatrix}$ |
| Rotation about the point $(0, 0)$ through $90°$ | $\begin{pmatrix} 0 & -1 \\ 1 & 0 \end{pmatrix}$ |
| Rotation about the point $(0, 0)$ through $180°$ | $\begin{pmatrix} -1 & 0 \\ 0 & -1 \end{pmatrix}$ |
| Rotation about the point $(0, 0)$ through $270°$ | $\begin{pmatrix} 0 & 1 \\ -1 & 0 \end{pmatrix}$ |
| Reflection in the line $x = 0$ | $\begin{pmatrix} -1 & 0 \\ 0 & 1 \end{pmatrix}$ |
| Reflection in the line $y = 0$ | $\begin{pmatrix} 1 & 0 \\ 0 & -1 \end{pmatrix}$ |
| Reflection in the line $x = y$ | $\begin{pmatrix} 0 & 1 \\ 1 & 0 \end{pmatrix}$ |
| Reflection in the line $x + y = 0$ | $\begin{pmatrix} 0 & -1 \\ -1 & 0 \end{pmatrix}$ |
| Enlargement centre $(0, 0)$, scale factor $k$ | $\begin{pmatrix} k & 0 \\ 0 & k \end{pmatrix}$ |
| Shear invariant line $y = 0$, $(0, 1) \rightarrow (k, 1)$ | $\begin{pmatrix} 1 & k \\ 0 & 1 \end{pmatrix}$ |
| Shear invariant line $x = 0$, $(1, 0) \rightarrow (1, k)$ | $\begin{pmatrix} 1 & 0 \\ k & 1 \end{pmatrix}$ |

To determine the matrix of a given transformation we observe that

$$\begin{pmatrix} a & b \\ c & d \end{pmatrix} \begin{pmatrix} 1 \\ 0 \end{pmatrix} = \begin{pmatrix} a \\ c \end{pmatrix} \qquad \begin{pmatrix} a & b \\ c & d \end{pmatrix} \begin{pmatrix} 0 \\ 1 \end{pmatrix} = \begin{pmatrix} b \\ d \end{pmatrix}.$$

Thus the first column of the matrix is the image of $\begin{pmatrix} 1 \\ 0 \end{pmatrix}$ under the transformation and the second column of the matrix is the image of $\begin{pmatrix} 0 \\ 1 \end{pmatrix}$ under the transformation; e.g. if a transformation maps the point $(1,0)$ onto the point $(2,1)$ and the point $(0,1)$ onto the point $(-3,6)$ we have

$$\begin{pmatrix} 1 \\ 0 \end{pmatrix} \rightarrow \begin{pmatrix} 2 \\ 1 \end{pmatrix}, \qquad \begin{pmatrix} 0 \\ 1 \end{pmatrix} \rightarrow \begin{pmatrix} -3 \\ 6 \end{pmatrix}.$$

Thus the matrix of the transformation is $\begin{pmatrix} 2 & -3 \\ 1 & 6 \end{pmatrix}$.

## EXERCISES

In this exercise the unit square $OABC$ has vertices $O(0,0)$, $A(1,0)$, $B(1,1)$ and $C(0,1)$.

1. Find the coordinates of the vertices of the image of the unit square $OABC$ under the transformations represented by the following matrices.

(a) $\begin{pmatrix} -3 & 0 \\ 0 & -3 \end{pmatrix}$,

(b) $\begin{pmatrix} 3 & 0 \\ 0 & 1 \end{pmatrix}$,

(c) $\begin{pmatrix} 2 & 1 \\ -1 & 0 \end{pmatrix}$,

(d) $\begin{pmatrix} 4 & -3 \\ 2 & 1 \end{pmatrix}$,

(e) $\begin{pmatrix} 1 & 0 \\ 0 & -1 \end{pmatrix}$,

(f) $\begin{pmatrix} 1 & 0 \\ 4 & 1 \end{pmatrix}$,

(g) $\begin{pmatrix} -1 & 3 \\ 1 & 2 \end{pmatrix}$,

(h) $\begin{pmatrix} 3 & 6 \\ 1 & 2 \end{pmatrix}$,

(i) $\begin{pmatrix} 1 & 0 \\ 0 & 1 \end{pmatrix}$.

Represent each of your results diagrammatically and, where possible, describe the transformation geometrically.

2. The image $A'B'C'$ of a triangle $ABC$ under a transformation

whose matrix is $\begin{pmatrix} 1 & 1 \\ -1 & 2 \end{pmatrix}$ has vertices $A'(3,6)$, $B'(1,5)$ and $C'(3,0)$. By finding the matrix of the inverse transformation, obtain the coordinates of the vertices of the triangle $ABC$.

3. State the single transformation equivalent to a half-turn about the point $(0,0)$ followed by a reflection in the line $x = 0$. Write down two matrices which correspond to these two transformations and check that their product corresponds to the combined transformation.

4. If a transformation **A** has matrix $\begin{pmatrix} 2 & 0 \\ 1 & -2 \end{pmatrix}$ and a transformation **B** has matrix $\begin{pmatrix} -1 & 2 \\ 0 & -2 \end{pmatrix}$, state the matrix corresponding to the transformation **AB**. Is the transformation **AB** the same as the transformation **BA**?

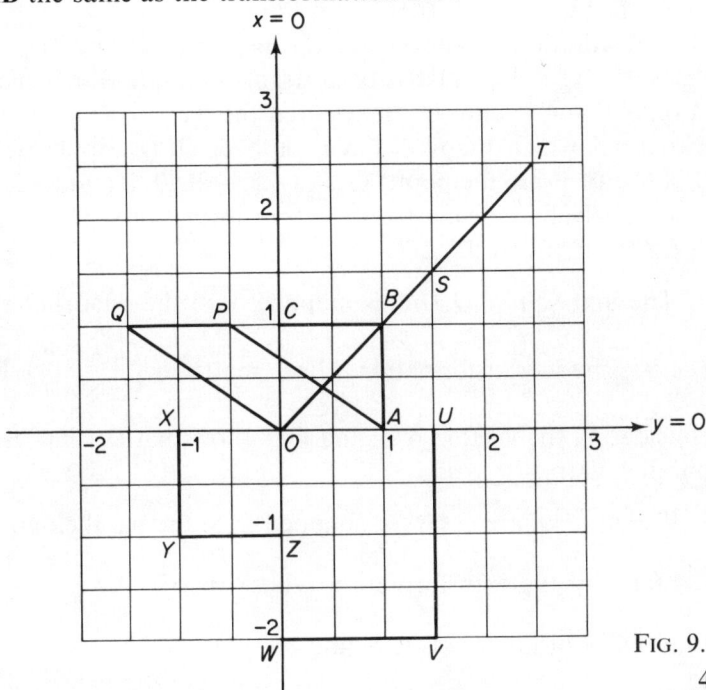

FIG. 9.2

5. Fig. 9.2 shows the image of the unit square under various transformations. Give the matrices of the transformations that map

(a) $OABC$ onto $OXYZ$,  (b) $OABC$ onto $OAPQ$,
(c) $OABC$ onto $OBST$,  (d) $OABC$ onto $OUVW$.

By finding the inverse of the matrix obtained in (d), give the matrix of the transformation that maps $OUVW$ onto $OABC$. Hence, find the matrix of the transformation that maps $OUVW$ onto $OAPQ$. Similarly, find the matrix of the transformation that maps $OUVW$ onto $OXYZ$.

Is it possible to find the matrix of a transformation which will map $OBST$ onto $OUVW$?

6. The triangle $OXY$ has vertices $O(0,0)$, $X(2, -1)$ and $Y(-1, 1)$. If $A$ is the point $(1, 0)$ and $C$ the point $(0, 1)$ give the matrix of the transformation which maps $A$ onto $X$ and $C$ onto $Y$. Hence, find the matrix of the transformation which maps $\triangle OXY$ onto $\triangle OAC$.

If in addition $P$ and $Q$ are the points $(1, 4)$ and $(1, -2)$ respectively, find the matrix of the transformation which maps $A$ onto $P$ and $C$ onto $Q$. Hence, find the matrix of the transformation which maps $\triangle OXY$ onto $\triangle OPQ$. Similarly, if $S, T, U$ and $V$ are the points $(3, 4)$, $(-1, -2)$, $(0, 4)$ and $(-3, 1)$ respectively, find the matrix of the transformation which maps $\triangle OST$ onto $\triangle OUV$.

7. The unit square $OABC$ is mapped onto the quadrilateral $OA'B'C'$ by a transformation whose matrix is $\begin{pmatrix} a & b \\ c & d \end{pmatrix}$. By considering the vectors $\mathbf{A'B'}$ and $\mathbf{OC'}$ show that $OA'B'C'$ is a parallelogram.

8. The unit square $OABC$ is mapped onto the parallelogram $OA'B'C'$ by the transformation whose matrix is $\begin{pmatrix} 3 & 1 \\ 4 & 2 \end{pmatrix}$.

Find the coordinates of $A', B'$ and $C'$.

The parallelogram $OA'B'C'$ is now transformed by a shear whose matrix is $\begin{pmatrix} 1 & -1\frac{1}{3} \\ 0 & 1 \end{pmatrix}$ followed by a shear whose matrix is $\begin{pmatrix} 1 & 0 \\ -1\frac{1}{2} & 1 \end{pmatrix}$. Find the area of the resulting figure and deduce the area of the parallelogram. By what factor has the area of the unit square changed under the original transformation?

9. By considering the matrix product $\begin{pmatrix} a & b \\ c & d \end{pmatrix}\begin{pmatrix} 0 \\ 0 \end{pmatrix}$ explain why a translation cannot be represented by a $2 \times 2$ matrix.

If the point $(x, y)$ is represented by the vector

$$\begin{pmatrix} x \\ y \\ 1 \end{pmatrix},$$

describe the transformation represented by the matrix

$$\begin{pmatrix} 1 & 0 & a \\ 0 & 1 & b \\ 0 & 0 & 1 \end{pmatrix}.$$

Describe the combination of transformations represented by the matrix

$$\begin{pmatrix} 0 & -1 & 2 \\ 1 & 0 & 3 \\ 0 & 0 & 1 \end{pmatrix}.$$

If this transformation is represented by

$$\begin{pmatrix} x \\ y \end{pmatrix} \rightarrow \mathbf{A}\begin{pmatrix} x \\ y \end{pmatrix} + \begin{pmatrix} c \\ d \end{pmatrix},$$

where $\mathbf{A}$ is a $2 \times 2$ matrix, find $\mathbf{A}$, $c$ and $d$.

# Chapter 10
# Graphical Representation of Equations

If $y$ is the image of $x$ under a function $f$ then the set of points whose coordinates are $(x, y)$, $\{(x, y): y = f(x)\}$, forms the graph of the function $f$, e.g. the graph of the function $x \to 2x - 3$ is drawn by substituting various values of $x$ into the equation $y = 2x - 3$ and calculating the corresponding values of $y$. The points so obtained are plotted and joined together by a smooth line to give the required graph. In general for a linear equation (i.e. an equation which can be expressed in the form $y = mx + c$ or $x = c$ where $m$ and $c$ are constants), the graph is a straight line.

If $(x_1, y_1)$ and $(x_2, y_2)$ are any two points on a straight line graph, the gradient of the line is defined to be $(y_2 - y_1)/(x_2 - x_1)$. This measures the steepness of the graph.

For the line $\{(x, y): y = mx + c\}$ the gradient is equal to $m$ and $c$ is the intercept on $x = 0$ (the value of $y$ when $x = 0$).

If the equation connecting $x$ and $y$ is non-linear, the graph of the function is not a straight line. Taking, for example, $y = x^2$, the following table shows corresponding values of $x$ and $y$:

| $x$ | $-3$ | $-2$ | $-1$ | 0 | 1 | 2 | 3 |
|---|---|---|---|---|---|---|---|
| $y$ | 9 | 4 | 1 | 0 | 1 | 4 | 9 |

and the graph of $y = x^2$ for these values of $x$ is shown in Fig. 10.1.

The graph of $y = f(x)$ divides the plane into two regions, one of these regions consisting of the points whose coordinates $(x, y)$ satisfy the inequality $y < f(x)$ and the other consisting of points whose coordinates $(x, y)$ satisfy the inequality $y > f(x)$. A region can be identified by choosing any point in the region and substituting its coordinates into each inequality to find the one they satisfy. For example, to represent the

44

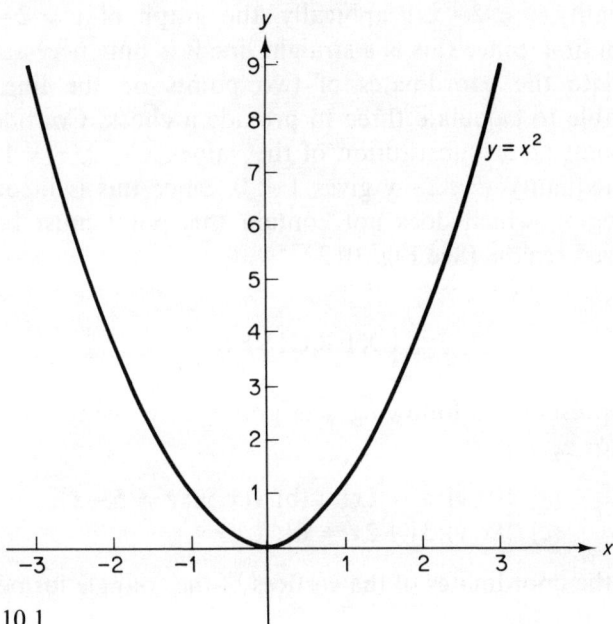

FIG. 10.1

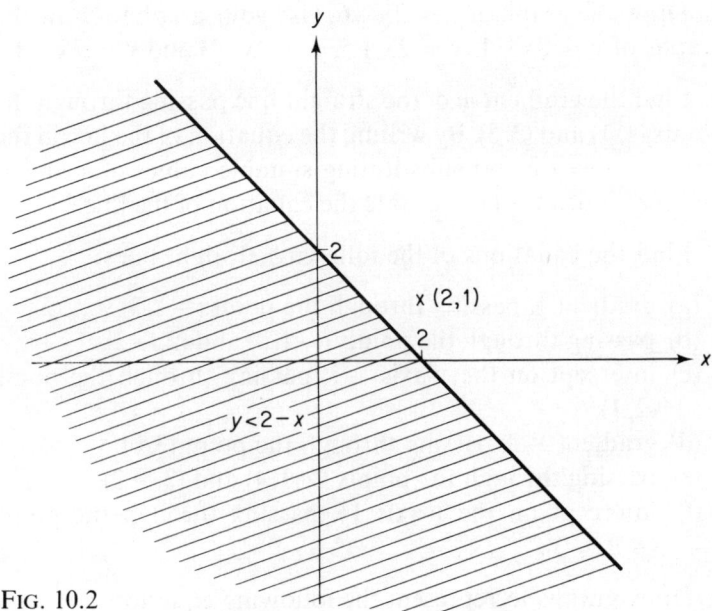

FIG. 10.2

inequality $y < 2-x$ graphically, the graph of $y = 2-x$ is drawn first. Since this is a straight line it is only necessary to calculate the coordinates of two points on the line, but desirable to calculate three to provide a check. Considering the point $(2, 1)$, substitution of the values $x = 2$, $y = 1$ into the inequality $y < 2-x$ gives $1 < 0$. Since this is incorrect, the region which does not contain this point must be the required region. (See Fig. 10.2.)

## EXERCISES

1. Represent the following sets graphically, using the same axes for each part.

(a) $\{(x, y): y = 2x\}$,  (b) $\{(x, y): y = 5-x\}$,
(c) $\{(x, y): 3y + 2x = 6\}$.

Give the coordinates of the vertices of the triangle formed by these three lines.

2. Draw the graph of $y = 2x + 6$. Use your graph to draw the graphs of $y = 2x - 1$, $y = 2x + 5$, $y = 2x + 3$ and $y = 2x - 4$.

3. Find the gradient $m$ of the straight line passing through the points $(1, 1)$ and $(3, 5)$. By writing the equation of the line in the form $y = mx + c$ and substituting suitable values of $x$ and $y$, find the value of $c$. Hence state the equation of the line.

4. Find the equations of the following straight lines:

(a) gradient $\frac{1}{2}$, passing through the point $(-1, 3)$,
(b) passing through the points $(-1, 3)$ and $(2, -2)$,
(c) intercept on the $y$-axis $-1$, passing through the point $(2, 1)$,
(d) gradient $-2$, passing through the point $(3, 0)$,
(e) passing through the points $(-1, 4)$ and $(3, -2)$,
(f) intercept on the $y$-axis $1\frac{1}{2}$, passing through the point $(-2, -3)$.

5. Draw graphs to represent the following equations:

46

(a) $y = \sqrt{x}$,　　　　(b) $xy = 12$,　　　　(c) $y = x^3$,

(d) $y = 2x^2 + 3$,　　　(e) $y = 2^x$,　　　　(f) $y = 1/(x+1)$.

6. Represent the following inequalities graphically.

(a) $x > 3$,　　　　　(b) $y > x+1$,　　　　(c) $x+2y < 6$,

(d) $2x - 3y > 4$,　　(e) $3(x+y) < 7$,　　　(f) $2x > 5(y-2)$,

(g) $y < x^2$,　　　　(h) $xy < 12$,　　　　(i) $x^2 < 0$.

7. The table shows the height of a balloon at various times.

| Time (p.m.) | 1.00 | 1.30 | 2.00 | 2.30 | 3.00 | 3.30 | 4.00 |
|---|---|---|---|---|---|---|---|
| Height (km) | 0 | 2.1 | 3.5 | 4.4 | 4.7 | 4.3 | 3.3 |

Draw a graph to represent this information, and use your graph to estimate

(a) the height of the balloon at 1.45 p.m.,

(b) the time when the balloon is 4 km high,

(c) the greatest height attained by the balloon in this period.

8. A football ground can hold 42 000 people. The number of people inside the ground before a match is shown in the table.

| Minutes before kick-off | 60 | 50 | 40 | 30 | 20 | 10 | 0 |
|---|---|---|---|---|---|---|---|
| Number of people in ground | 600 | 2000 | 4200 | 7500 | 12 800 | 28 400 | 34 700 |

Represent this information graphically and use your graph to find

(a) when the ground is half full,

(b) the number of people who arrive within five minutes of the kick-off.

# Chapter 11
# Equations and Inequalities

*Equations*

An equation gives information about an unknown number from which the number must be found. Common sense must be used when solving equations. A vital rule which must be followed is that any operation performed on one side of the equation must be performed on the other, e.g.

(i) $$2x - 7 = 4$$
Adding 7 to each side, $$2x = 11$$
Dividing each side by 2, $$x = 5\tfrac{1}{2}.$$

This equation can be considered in terms of the function

$$f : x \to 2x - 7.$$

We are given that $f(x) = 4$, so $f^{-1}(4) = x$. The inverse function is

$$f^{-1} : x \to (x + 7)/2.$$

Thus $$x = f^{-1}(4) = 5\tfrac{1}{2}.$$

(ii) $$3x + 2 = 9 - 2x$$
Adding $2x$ to each side, $$5x + 2 = 9$$
Subtracting 2 from each side, $$5x = 7$$
Dividing each side by 5, $$x = 1\tfrac{2}{5}.$$

(iii) $$(2x + 1)/(x + 1) = 6$$
Multiplying each side by $x + 1$, $$2x + 1 = 6(x + 1)$$
Clearing brackets, $$2x + 1 = 6x + 6$$
Subtracting $2x$ from each side, $$1 = 4x + 6$$
Subtracting 6 from each side, $$-5 = 4x$$
Dividing each side by 4, $$-5/4 = x$$
$$x = -1\tfrac{1}{4}.$$

(iv) $$(x-3)(2x-3) = 0.$$

In solving equations of this type we use the fact that if $a$ and $b$ are numbers,

$$ab = 0 \Leftrightarrow a = 0 \text{ or } b = 0$$

Thus $(x-3)(2x-3) = 0 \Rightarrow x-3 = 0 \text{ or } 2x-3 = 0$

$$\Rightarrow x = 3 \text{ or } x = 1\tfrac{1}{2}.$$

*Inequalities*

Inequalities are solved in the same way as equations except that whenever both sides are multiplied or divided by a negative number the inequality sign must be reversed, e.g.

(v) $$2x-7 < 4$$

As in example (i) $\qquad\qquad x < 5\tfrac{1}{2}$

The solution set is $\{x : x < 5\tfrac{1}{2}\}$.

(vi) $$2-5x \leqslant 6$$

Subtracting 2 from each side, $\quad -5x \leqslant 4$

Dividing each side by $-5$, $\qquad x \geqslant -4/5$

The solution set is $\{x : x \geqslant -4/5\}$.

(vii) $$2(2x+1) > 3-(2-x)$$

Clearing brackets, $\qquad\qquad 4x+2 > 3-2+x$

i.e. $\qquad\qquad\qquad\qquad 4x+2 > 1+x$

Subtracting $x$ from each side, $\;\; 3x+2 > 1$

Subtracting 2 from each side, $\qquad 3x > -1$

Dividing each side by 3, $\qquad\qquad x > -\tfrac{1}{3}$

The solution set is $\{x : x > -\tfrac{1}{3}\}$.

(viii) $$(x+1)(2x-5) \leqslant 0$$

In solving inequalities of this kind we use the fact that if $a$ and $b$ are numbers

$$ab < 0 \Leftrightarrow a \text{ and } b \text{ have opposite signs}$$
$$ab > 0 \Leftrightarrow a \text{ and } b \text{ have the same sign}$$

Fig. 11.1 shows the regions of the number line for which

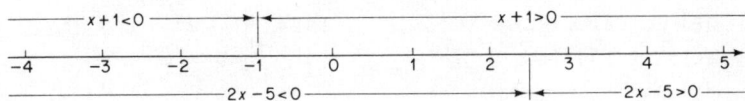

FIG. 11.1

$x+1$ and $2x-5$ are positive and negative. Thus $x+1$ and $2x-5$ have opposite signs in the region $-1 < x < 2\frac{1}{2}$, and so

$$(x+1)(2x-5) \leqslant 0 \Leftrightarrow -1 \leqslant x \leqslant 2\frac{1}{2}.$$

## EXERCISES

1. Solve the following equations.

(a) $2x+7 = 3$,
(b) $3-4x = 8$,
(c) $2-2x = 8$,
(d) $3(2+x) = 4x$,
(e) $2-3(x-1) = 6$,
(f) $3(x+1) = 4(2x-3)$,
(g) $1/(x+3) = 2$,
(h) $(x+3)/(2x-1) = 1$,
(i) $\frac{1}{3}x-2 = \frac{1}{4}x$,
(j) $\frac{2}{3}(x+5) = 1$,
(k) $(2x+1)/3 = 4$,
(l) $2-(2x-3) = 6$,
(m) $6 = 4-3x$,
(n) $5x-2 = 4-3x$,
(o) $2(1-x) = 3$,
(p) $3-(6-x) = 2x$.

2. Find the solution set of each of the following inequalities.

(a) $3x-5 > 1$,
(b) $1-2x \leqslant 4$,
(c) $3-(1-2x) < 4$,
(d) $x+6 \geqslant 5x$,
(e) $3(2+x) < 2x-3$,
(f) $4(2x+3) > 2(3x+7)$,
(g) $\frac{2}{3}x \geqslant 1-\frac{1}{2}x$,
(h) $2(x+1) > 2x+5$,
(i) $4-2(3x-1) > 3(2x-3)$,
(j) $3x+6 \leqslant 3(x+2)$.

3. Find the inverse of the function $f: x \to 3(2x-1)$. Use the inverse function to solve the equations

(a) $3(2x-1) = 5$,
(b) $3(2x-1) = -2$,
(c) $6x-3 = 4$,
(d) $2x-1 = 2\frac{1}{3}$.

Use a similar technique to solve the equations

(e) $5/(2x+1) = 2$,
(f) $5/(2x+1) = -3$.

4. State the values of $x$ satisfying each of the following equations.

(a) $(x-1)(x-2) = 0$,
(b) $(x+1)(2x-1) = 0$,
(c) $x(x+1) = 0$,
(d) $x^2-9 = 0$,
(e) $5x^2+2x = 0$.

5. Give the solution set of each of the following inequalities, and represent your solution on the number line.

(a) $(x-1)(x+2) \leqslant 0$,

(b) $(x+1)(2x-1) > 0$,

(c) $x(2x-3) < 0$,

(d) $3x^2 - 2x \geqslant 0$,

(e) $(2x-1)^2 > 0$.

# Chapter 12
# Simultaneous Equations

Two equations involving two unknowns $x$ and $y$ have a simultaneous solution if it is possible to find a value of $x$ and a value of $y$ which satisfy both equations. The following methods of solution are available.

(a) *By elimination*   The object here is to find a combination of the two equations which gives a single equation containing one unknown only, e.g.

$$2x - y = 5 \tag{1}$$
$$3x + 2y = 4 \tag{2}$$

Multiplying equation (1) by 2 gives

$$4x - 2y = 10.$$

Adding this equation to equation (2) we obtain

$$7x = 14.$$
Hence $\qquad x = 2.$

Substituting $x = 2$ in equation (1) gives

$$y = -1.$$

Note that $x = 2$, $y = -1$ satisfies equation (2).

(b) *Using matrices*   The simultaneous equations

$$ax + by = p$$
$$cx + dy = q$$

can be written in matrix form

$$\begin{pmatrix} a & b \\ c & d \end{pmatrix} \begin{pmatrix} x \\ y \end{pmatrix} = \begin{pmatrix} p \\ q \end{pmatrix}.$$

This single matrix equation can be solved by pre-multiplying each side of the equation by the inverse of the matrix $\begin{pmatrix} a & b \\ c & d \end{pmatrix}$.

In practice it is simpler to pre-multiply each side by the matrix $\begin{pmatrix} d & -b \\ -c & a \end{pmatrix}$, e.g.

$$x + 4y = 10$$
$$2x + 3y = 5$$

$$\begin{pmatrix} 1 & 4 \\ 2 & 3 \end{pmatrix} \begin{pmatrix} x \\ y \end{pmatrix} = \begin{pmatrix} 10 \\ 5 \end{pmatrix}.$$

Pre-multiplying by $\begin{pmatrix} 3 & -4 \\ -2 & 1 \end{pmatrix}$,

$$\begin{pmatrix} 3 & -4 \\ -2 & 1 \end{pmatrix} \begin{pmatrix} 1 & 4 \\ 2 & 3 \end{pmatrix} \begin{pmatrix} x \\ y \end{pmatrix} = \begin{pmatrix} 3 & -4 \\ -2 & 1 \end{pmatrix} \begin{pmatrix} 10 \\ 5 \end{pmatrix}$$

$$\begin{pmatrix} -5 & 0 \\ 0 & -5 \end{pmatrix} \begin{pmatrix} x \\ y \end{pmatrix} = \begin{pmatrix} 10 \\ -15 \end{pmatrix}$$

$$\begin{pmatrix} -5x \\ -5y \end{pmatrix} = \begin{pmatrix} 10 \\ -15 \end{pmatrix}$$

$$-5x = 10$$
$$-5y = -15$$

giving $x = -2$ and $y = 3$.

(c) *Graphically* The solution of any two simultaneous equations can be found by obtaining the coordinates of the points where the graphs of the two equations intersect. This method, unlike the previous two, is applicable to both linear and non-linear equations, e.g.

$$y = x^2, \qquad y = x + 2$$

From Fig. 12.1 the solutions are $x = 2$, $y = 4$ and $x = -1$, $y = 1$.

Simultaneous linear equations can have no solution, one solution or infinitely many solutions. The three cases can be demonstrated graphically.

In Fig. 12.2, the two lines intersect at one point and the

53

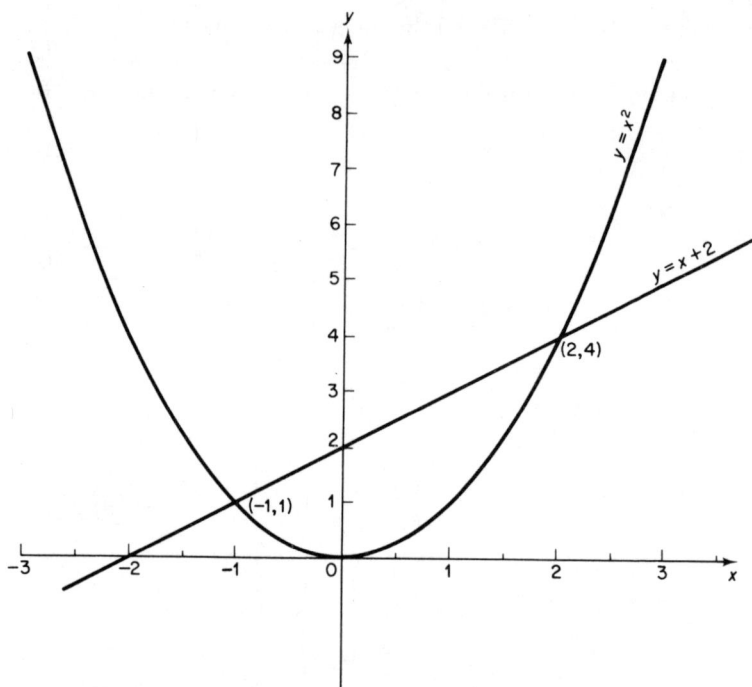

FIG. 12.1

equations have one solution. In Fig. 12.3, the two lines are parallel and the equations have no solution. In Fig. 12.4, the two equations represent the same line and hence any point on

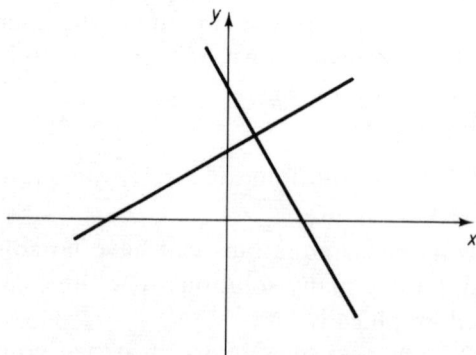

FIG. 12.2

54

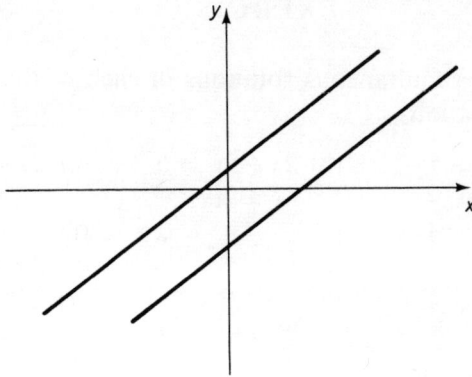

FIG. 12.3

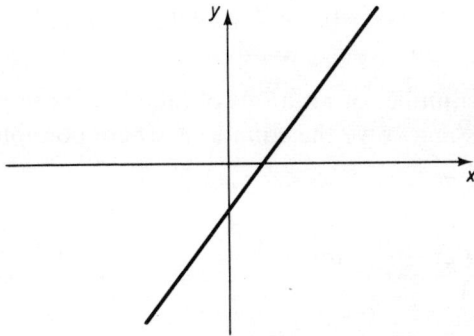

FIG. 12.4

this line satisfies both equations. The equations have in-finitely many solutions.

If the two equations are

$$ax + by = p$$
$$cx + dy = q$$

and $\Delta = ad - bc$, then

case (a) corresponds to $\Delta \neq 0$,
case (b) corresponds to $\Delta = 0$, one equation not being a multiple of the other,
case (c) corresponds to $\Delta = 0$, one equation being a multiple of the other.

55

## EXERCISES

1. Give the simultaneous solutions of each of the following pairs of equations.

(a) $x+2y = 5$
   $5x+2y = 9$

(b) $2x+5y = 5$
   $6y = 12$

(c) $2x-y = 10$
   $3x+2y = 1$

(d) $2x+7y = 4$
   $3x+8y = 5$

(e) $3x-y = 2$
   $5x-4y = 5$

(f) $3x-y = 22$
   $5x-2y = 3$

(g) $4x+9y = 3$
   $3x+4y = 1$

(h) $y = 3x+2$
   $2y+5x = 15$

(i) $x+2y = 1\frac{1}{3}$
   $3y-x = 1\frac{4}{9}$

(j) $5x-6y+9 = 0$
   $x-y = 2$

(k) $5x+3y = 8$
   $3x+5y = -2$

(l) $y = x^2$
   $y = 3x-2$

(m) $xy = 8$
   $2x+3y = 19$

(n) $y = x^3+1$
   $x+4y=5.$

2. State the number of solutions of each of the following pairs of equations and solve the equations where possible.

(a) $4x-y = 7$
   $x+4y = 6$

(b) $x+2y = 1$
   $2x+4y = 2$

(c) $2x-y = 3$
   $4x-2y = 7$

(d) $y = x^2+2$
   $y = x-1$

(e) $y = x^3$
   $y = x$

(f) $y = 1-3x$
   $x = 3-\frac{1}{3}y.$

3. Evaluate the matrix product

$$\begin{pmatrix} -5 & 1 & 3 \\ 7 & 1 & -5 \\ 1 & -1 & 1 \end{pmatrix} \begin{pmatrix} 1 & 1 & 2 \\ 3 & 2 & 1 \\ 2 & 1 & 3 \end{pmatrix}.$$

Use your answer to solve the simultaneous equations

$$x+y+2z = 1$$
$$3x+2y+z = 7$$
$$2x+y+3z = 2.$$

4. The straight line whose equation is of the form $y = ax+b$, where $a$ and $b$ are numbers, passes through the points $(2, 1)$ and $(-3, -9)$. Form two equations involving $a$ and $b$ and by finding the simultaneous solution write down the equation of the line.

5. The translation **P** has column vector $\begin{pmatrix} 2 \\ -5 \end{pmatrix}$ and the translation **Q** has column vector $\begin{pmatrix} -3 \\ 2 \end{pmatrix}$. If $\mathbf{P}^x\mathbf{Q}^y = \mathbf{S}$, where **S** is a translation with column vector $\begin{pmatrix} 0 \\ -11 \end{pmatrix}$, find $x$ and $y$.

6. A car park which can hold a maximum of five hundred cars also caters for coaches, a coach occupying five times the space of a car. The cost for a car is 25p per day and for a coach 50p per day. If on a certain day the car park is initially empty and £110 is taken at the entrance, find the number of cars and the number of coaches, assuming that each space is filled and that each space has been used only once.

7. A breakfast-cereal manufacturer offers bowls at the reduced price of 10p and two packet tops, and milk jugs at the reduced price of 12p and three packet tops. One week the manufacturer receives 1679 packet tops and £77·80. How many bowls and milk jugs are required that week?

8. I ask my milkman to deliver 'yellow' top milk but because I am near the end of his round he has to leave me 'pink' top if he has no 'yellow' top left. A bottle of 'yellow' top milk costs 5p and a bottle of 'pink' top costs $5\frac{1}{2}$p. At the end of a certain week I am given a bill for £1·18 and doubt its validity. I know that my usual order of three bottles of milk per weekday and four bottles on each of Saturday and Sunday has been delivered. Could the bill be correct and if so how many bottles of 'pink' top milk have been delivered during the week?

# Chapter 13
# Linear Programming

The graphical method for finding the solution set of a single inequality outlined in Chapter 11 can be extended to give the solution set of two or more simultaneous inequalities. When solving two or more simultaneous inequalities it is advisable to obtain the solution set of each inequality separately, shading in the region which is not required. After all the inequalities have been solved, the simultaneous solution is represented by the region left unshaded. The simultaneous solution of the inequalities

$$x + y < 10$$
$$y < x + 4$$

is shown in Fig. 13.1. The horizontal shading removes points which do not satisfy the inequality $y < x + 4$ and the vertical shading removes points which do not satisfy the inequality $x + y < 10$. The only points satisfying both inequalities lie in the unshaded region.

If the inequality is of the type involving $\leqslant$ or $\geqslant$ then the boundary line is part of the solution set. In Fig. 13.1 the lines $x + y = 10$ and $y = x + 4$ are not part of the solution set and it is sometimes useful to use broken lines in this case.

In linear programming it is necessary to formulate inequalities, usually referring to some practical situation, and apply this technique to their solution. For example, I wish to knit a scarf using red and blue wool. I do not wish to use more than three hanks of red wool or more than four hanks of wool altogether. Red wool costs 10p per hank and blue wool costs 20p per hank. I have 60p to spend on the scarf. Represent this information graphically.

Assume I use $x$ hanks of blue wool and $y$ hanks of red wool. Then

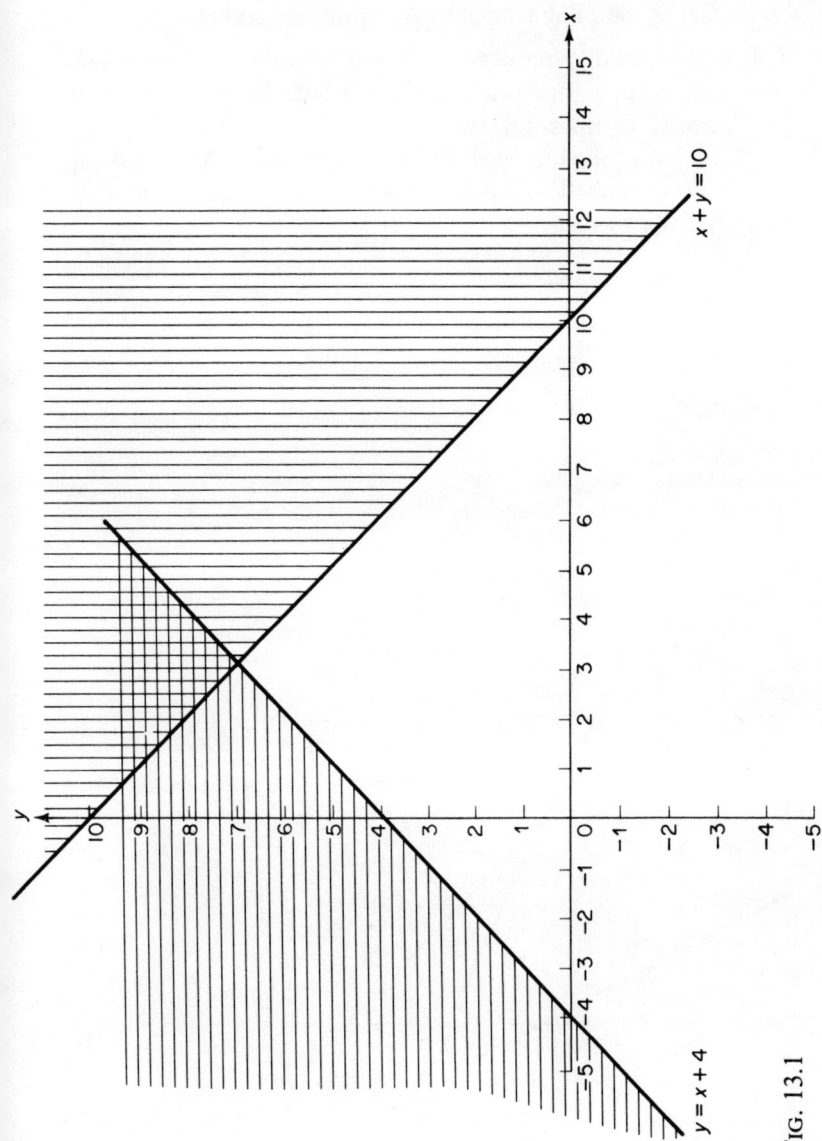

$x + y = 10$

$y = x + 4$

Fig. 13.1

59

$y \leqslant 3$   (I do not use more than 3 hanks of red wool.)
$x + y \leqslant 4$   (The total number of hanks must not exceed 4.)
$20x + 10y \leqslant 60$ (The total cost must not exceed 60p.)

The last inequality reduces to $2x + y \leqslant 6$. Fig 13.2 shows the solutions of these inequalities. The unshaded region represents their simultaneous solution.

If one hank of blue wool knits into 20 cm of scarf and one hank of red wool knits into 40 cm of scarf, how long is the longest scarf I can knit?

FIG. 13.2

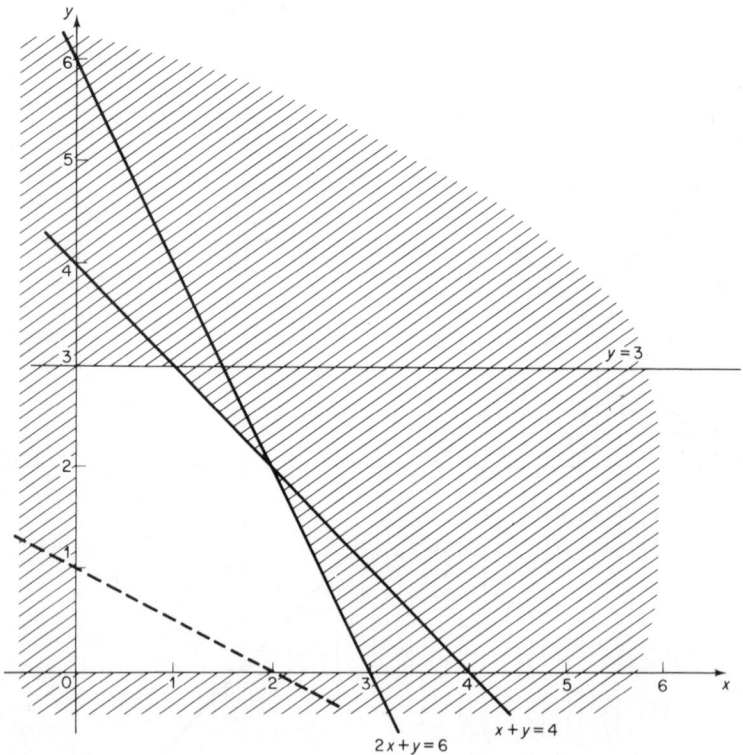

Length of scarf $= (20x + 40y)$ cm. If we take $20x + 40y = l$, all the lines $20x + 40y = l$ for varying values of $l$ are parallel. Taking $l = 40$ we obtain the broken line in Fig. 13.2. The line parallel to this broken line which is farthest from the origin and passes through a point of the solution set has equation $20x + 40y = L$, where $L$ is the length of the longest scarf I can knit. Here this line passes through the point $(1, 3)$ of the solution set, so the longest scarf I can knit uses 1 hank of blue wool and three hanks of red wool, and has length 140 cm.

## EXERCISES

1. Draw the graphs of the equations $x + y = 6$, $y = x - 2$, $y = 2x$ and $y = 1$. Indicate on separate graphs the regions corresponding to the following sets:

    (a) $\{(x, y) : x + y > 6 \text{ and } y < 2x\}$,
    (b) $\{(x, y) : y \geqslant 1, x + y \leqslant 6 \text{ and } y \geqslant 2x\}$,
    (c) $\{(x, y) : y > 1, x + y \leqslant 6 \text{ and } y \leqslant x - 2\}$,
    (d) $\{(x, y) : y \geqslant 1, y < 2x, x + y \leqslant 6 \text{ and } y > x - 2\}$.

2. Represent graphically the set of points satisfying both the inequalities

$$y > x + 1$$
$$2y < x + 6.$$

3. Represent graphically

$$\{(x, y) : y \geqslant 0, y \leqslant x, x \leqslant 4 \text{ and } x + y \geqslant 2\}.$$

Subject to these conditions find the maximum and minimum values of (a) $2y + x$, (b) $y + 2x$.

4. A car-hire firm buys a number of new cars of two types. Type $A$ cost £800 each and £6 per week to run, whilst type $B$ cost £1200 each and £5 per week to run. At the end of one year the cars are sold for £500 and £800 respectively. The firm intends to spend a maximum of £12 000 and hopes to get at

least £6000 at the end of the year on selling these cars. If the firm does not wish to spend more than £60 per week on maintenance, find the possibilities open to the firm.

5. A schoolboy arrives home at 4.30 p.m., spends half an hour eating and goes to bed between 9.00 p.m. and 10.00 p.m. The rest of the evening he spends on homework and his hobbies. He likes to spend more time on hobbies than homework, but has to spend at least two hours on his homework. If time appears to pass twice as quickly when he is enjoying his hobbies as when he is doing his homework, how should he arrange his time to

  (a) make the evening pass as quickly as possible,
  (b) make the evening appear to last as long as possible?

6. A fresh-fruit drink is made solely from the juices of oranges and lemons. One orange gives an average 8 cm$^3$ of juice and one lemon gives 5 cm$^3$ of juice. Oranges cost $2\frac{1}{2}$p each and lemons cost 2p each, and it is required to make at least 80 cm$^3$ of the drink at a cost of not more than 60p. If, for a satisfactory drink, the number of oranges used must exceed the number of lemons but be less than twice the number of lemons, show graphically the region corresponding to the numbers of oranges and lemons which will suffice.

7. A builder purchases a plot of land of area 30 000 m$^2$ on which he proposes to build detached and semi-detached houses. A semi-detached house requires an area of 300 m$^2$ and costs £4000 to build. A detached house needs an area of 500 m$^2$ and costs £5000 to build. The builder has £350 000 available to spend on the houses and has firm orders for fifteen detached and twenty semi-detached houses. He is required to build at least as many semi-detached houses as detached ones. Represent this information graphically, and estimate how many houses of each type the builder should build to gain maximum profit if he sells each detached house for £6000 and each semi-detached house for £4700.

8. A manufacturer makes cylindrical tins of volume at least 110 cm$^3$, subject to the restrictions that the base radius of the tins must not exceed 5 cm and must not be less than the height. Represent graphically these restrictions and deduce

    (a) the greatest possible volume of a tin,
    (b) the least possible height of a tin,
    (c) the least possible radius of a tin.

# Chapter 14
# Trigonometry

The position of a point $P$ is described by polar coordinates when the distance of $P$ from a fixed point $O$ and the angle made by $OP$ with a fixed line $OX$ are given. In Fig. 14.1 the polar coordinates of $P$ are $(r, \theta)$. If $Q$ is a point on $OX$ such that $O\hat{Q}P = 90°$, then by definition

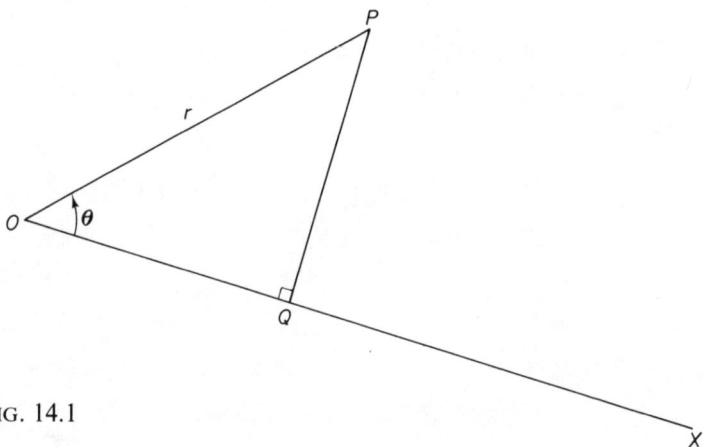

FIG. 14.1

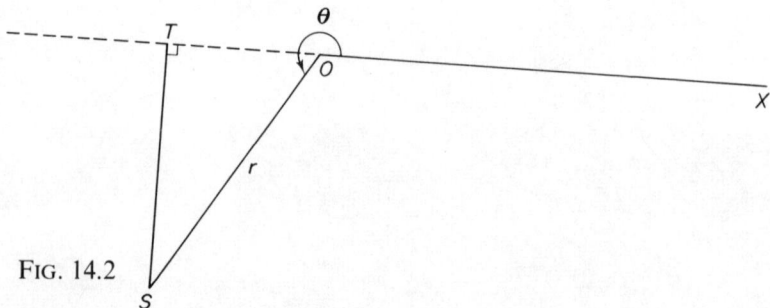

FIG. 14.2

64

$$OQ = r\cos\theta \qquad \text{(or } \cos\theta = OQ/r\text{)}$$
$$QP = r\sin\theta \qquad \text{(or } \sin\theta = QP/r\text{)}$$
$$PQ = OQ\tan\theta \qquad \text{(or } \tan\theta = PQ/OQ\text{)}.$$

These definitions apply for all values of $\theta$. In Fig. 14.2, $OT = r\cos\theta$ and $ST = r\sin\theta$. (Here $\cos\theta$ and $\sin\theta$ are both negative.)

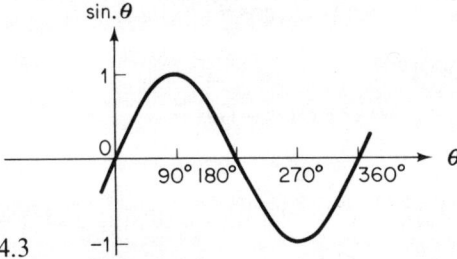

FIG. 14.3

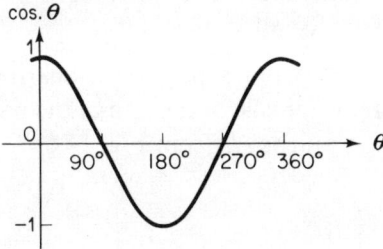

FIG. 14.4

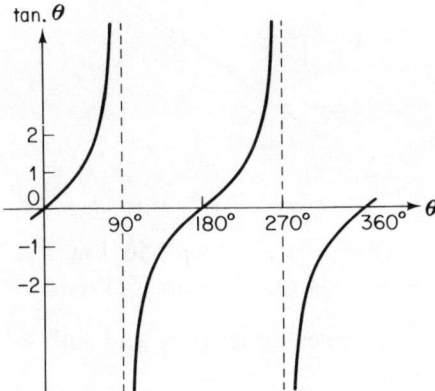

FIG. 14.5

65

To find the sine, cosine or tangent of an angle between $0°$ and $90°$, tables are used. For angles outside this range the symmetries of the graphs of these functions are useful (see Figs. 14.3, 4 and 5).

From the definition of tangent (Fig. 14.1),

$$\tan \theta = \frac{PQ}{OQ} = \frac{r \sin \theta}{r \cos \theta} = \frac{\sin \theta}{\cos \theta}.$$

By Pythagoras' theorem,

$$OP^2 = OQ^2 + QP^2.$$

Hence

$$r^2 = r^2 \cos^2 \theta + r^2 \sin^2 \theta \quad \text{(by definition of sine and cosine)}$$

$$\sin^2 \theta + \cos^2 \theta = 1.$$

Note that these results could also have been obtained from Fig. 14.2 and are true for all values of $\theta$.

*Example* A ship sails from a port on a bearing $056°$ for a distance of 8 km. How far east and north of the port is the ship after this stage of its journey? From Fig. 14.6,

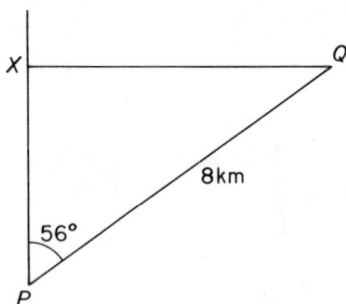

Fig. 14.6

distance north $= PX = 8 \cos 56° \text{ km} = 4 \cdot 77 \text{ km}$
distance east $\;= QX = 8 \sin 56° \text{ km} = 6 \cdot 63 \text{ km}.$

The ship now continues its journey and sails a further 5 km on a bearing of $112°$.

66

Total distance north $= (8 \cos 56° + 5 \cos 112°)$ km
$$= (4·77 - 1·88) \text{ km}$$
$$= 2·89 \text{ km.}$$

Total distance east $= (8 \sin 56° + 5 \sin 112°)$ km
$$= (6·63 + 4·63) \text{ km}$$
$$= 11·26 \text{ km.}$$

If the bearing of the ship from the port is now $\theta$,

$$\tan \theta = 11·26/2·8 = 4·02$$

Hence $\qquad\qquad \theta = 76°.$

## EXERCISES

1. A square trap-door of side 800 cm is hinged 300 cm from a vertical wall. Through what angle can the trap-door be turned when opened upwards?

2. A ladder 6 m long is to be placed against a wall to reach a window 4 m high. How far from the base of the wall should the ladder be placed and what angle will the ladder make with the horizontal?

   If the bottom of the ladder is left in this position and the ladder is extended to reach a window 6 m high, by how much must the ladder be extended?

3. In a triangle $ABC$, $AB = 14$ cm, $A\hat{B}C = 37°$ and $A\hat{C}B = 42°$. By finding the length of the perpendicular from $A$ to $BC$ calculate the length of the side $AC$.

4. A ship leaves port on a bearing of 032° and sails for 15 km. It then changes course to a bearing of 123° and continues in this direction for 10 km. At the end of this stage of its journey calculate

   (a) how far north of the port the ship is,
   (b) how far east of the port the ship is,
   (c) the distance of the ship from port,
   (d) the bearing of the port from the ship.

67

5. When standing 34 m from the base of a tower I measure the angle of elevation of the top of the tower to be 42°. How high is the tower? If I now walk a further 10 m away from the tower what is the new angle of elevation of the top of the tower?

6. A lighthouse is 10 km from a port on a bearing of 23° from the port. If a boat sails on a bearing of 38° from the port how close to the lighthouse will it pass?

If the area within 2 km of the lighthouse is dangerous to shipping, on what bearings can a ship sail to get as close as possible to the lighthouse?

7. On a big wheel at a fair the height $h$ m of my seat above the ground $t$ sec after the wheel begins to rotate is given by

$$h = 8 - 6 \cos 10t.$$

Draw a graph to show my height above ground level during the first complete revolution of the wheel. Hence, or otherwise, state

    (a) the greatest height I attain,
    (b) the radius of the wheel,
    (c) the time taken for one complete revolution,
    (d) the time for which I am more than 10 m above ground level.

8. A submarine is submerged with its periscope above the level of the water. As a result of the waves passing over the submarine the length of periscope above water level, $l$ m, varies according to the formula

$$l = 1 + \tfrac{1}{2} \cos 60t$$

where $t$ is the time in seconds after raising the periscope. Draw a graph to show the length of periscope above water level over the first six seconds. If clear vision is only possible when at least 0·75 m of periscope is above water level, for what fraction of the total time is vision obscured?

9. Fig. 14·7 shows the plan of a harbour with a boat at $P$ 50 m away from the outer wall. The harbour is 50 m wide and the
68

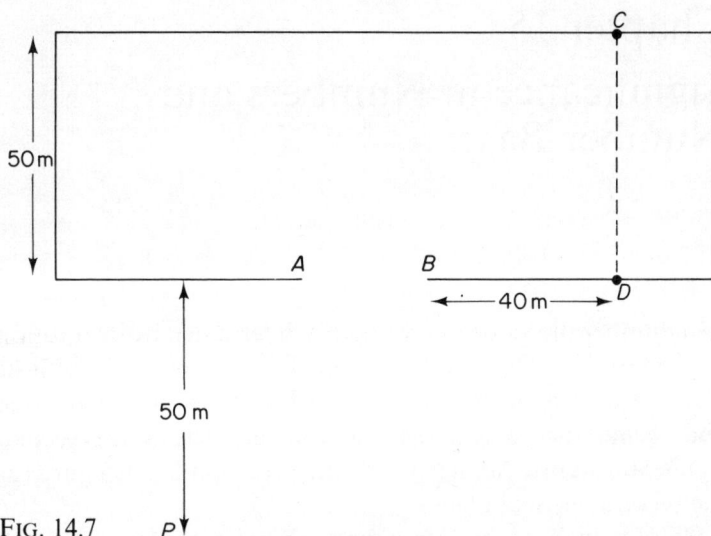

FIG. 14.7

entrance *AB* is 10 m wide. The bearing of *B* from the boat is 045°. Between which bearings must the boat sail if it is to pass through the entrance to the harbour?

The boat aims to tie up at the point *C* on the inner wall (see Fig. 14.7). On what bearing must the boat sail to head straight towards *C*?

10. (i) Convert the following polar coordinates to cartesian coordinates:

   (a) $(5, 30°)$,   (b) $(8, 208°)$,   (c) $(6, 143°)$,   (d) $(4, 292°)$.

   (ii) Convert the following cartesian coordinates to polar coordinates:

   (a) $(4, 5)$,   (b) $(-2, 4)$,   (c) $(-3, -6)$,   (d) $(3, -5)$.

11. Without using trigonometrical tables

   (a) find $\cos \theta$ and $\tan \theta$ if $\sin \theta = 0.8$ $(90° < \theta < 270°)$,
   (b) find $\sin \theta$ and $\tan \theta$ if $\cos \theta = -0.5$ $(180° < \theta < 360°)$,
   (c) find $\sin \theta$ and $\cos \theta$ if $\tan \theta = 1.4$ $(0° < \theta < 180°)$.

69

# Chapter 15
# Significance in Numbers and Number Bases

In a number the value of any digit is determined by its position in the number. The significance of the digit increases with its place value, e.g. in the number 376·284, 7 has place value ten and, being the second digit in the number, is the second significant figure; similarly 8 is the fifth significant figure (or the second decimal place).

Zeros which occur in a number merely to determine the place value of other digits are not significant, e.g. the number 76 000 is given to two significant figures (2 s.f.), whereas in the number 76 000·0 the zero in the decimal place indicates that the number is given correct to one decimal place (1 d.p.) and hence the zeros are significant.

Giving a number to $n$ significant figures (or decimal places) involves finding a number with $n$ significant figures (or decimal places) as close as possible to the given number, e.g.

$$376·234 = 380 \quad \text{(to 2 s.f.),}$$
$$376·234 = 376·2 \quad \text{(to 1 d.p.).}$$

In the decimal system of numbers, place values are determined on powers of ten. Ten is called the base of the system. In other number systems different numbers may be used for the base.

In base ten, $234 = 2 \times 10^2 + 3 \times 10 + 4$.

In base six, $234 = 2 \times 6^2 + 3 \times 6 + 4 \quad (= 94 \text{ in base ten}).$

234 in base six is written $234_6$.

To convert a number in base ten to another base, the number can be broken down into multiples of the powers of the base, e.g.

$$56_{10} = 1 \times 36 + 3 \times 6 + 2 = 132_6.$$

In practice successive division is usually simpler, e.g.

$$
\begin{array}{r|l}
6 & 56 \\
\hline
6 & 9 \quad \text{r. } 2 \\
6 & 1 \quad \text{r. } 3 \\
\hline
& 0 \quad \text{r. } 1
\end{array}
$$

Thus $56 = 132_6.$

Arithmetical operations in any base are performed in exactly the same way as those in base ten, except that the base number is carried rather than ten, e.g.

(a) $\quad 235_7 +$
$\phantom{(a)}\quad 633_7$
$\phantom{(a)}\quad \overline{1201_7}$

(c) $412_5 -$
$\phantom{(c)}\ 241_5$
$\phantom{(c)}\ \overline{121_5}$

(b) $\quad 2102_3 \times$
$\phantom{(b)}\qquad 2_3$
$\phantom{(b)}\quad \overline{11211_3}$

(d) $13_6)\overline{1202_6}$ $\quad 32_6$
$\phantom{(d)}\qquad 111_6$
$\phantom{(d)}\qquad \overline{\ 32_6}$
$\phantom{(d)}\qquad \ \ 32_6$

Numbers to base two are referred to as binary numbers.

The principle of decimal fractions can be applied to other bases, e.g.

(a) $21 \cdot 37_{10} = 2 \times 10 + 1 + \frac{3}{10} + \frac{7}{100},$
(b) $21 \cdot 37_8 \ = 2 \times 8 + 1 + \frac{3}{8} + \frac{7}{64}.$

## EXERCISES

Complete the following tables:

1.

| Number | to 1 s.f. | to 3 s.f. | to 1 d.p. |
|---|---|---|---|
| 237·26 | 200 | 237 | 237·3 |
| 1·7491 | | | |
| 0·07359 | | | |
| 2367 | | | |
| 11·447 | | | |
| 0·003288 | | | |
| 2761·89 | | | |
| 0·000321 | | | |

2.

| base 10 | base 2 | base 5 | base 7 |
|---|---|---|---|
| 23 | | | |
| | 111011 | | |
| 96 | | | |
| | | | 100 |
| | | 321 | |
| | 11 | | |
| | | | 236 |
| 182 | | | |
| | | 10 | |

72

3. Calculate

  (a) $110111_2 + 1110_2$,      (b) $235_7 - 62_7$,
  (c) $1221_3 \times 2_3$,      (d) $3302_5 \div 4_5$,
  (e) $932_{12} + 719_{12}$,      (f) $1111_6 - 555_6$,
  (g) $346_8 \times 37_8$,      (h) $10000111_2 \div 101_2$.

4. State the base of the following calculations:

  (a) $123 + 23 = 201$,      (b) $273 - 125 = 147$,
  (c) $231 \times 3 = 2013$,      (d) $1274 \div 5 = 214$.

5. (a) Convert $0 \cdot 75_{10}$ to   (i) base 2   (ii) base 12.
  (b) Convert $23 \cdot 21_4$ to base 10.
  (c) Convert $5_6/12_6$ to a base 6 'decimal'.
  (d) Convert $0 \cdot 3\dot{2}_5$ to a base 5 fraction.

6. After a lesson on number bases, a boy has a nightmare in which he dreams that he is 22 years old and was born on 40/15/13020. If in reality the boy is 14 years old, in what number base was his dream, and when was he born?

7. Jim, Dick and Dave are discussing their weights. Jim claims to weigh 322 kg, Dick to weigh 47 kg and Dave to weigh 202 kg. Unfortunately each of them has given his weight to a different base. I know that Jim weighs more than Dick who weighs more than Dave, and that no one weighs more than $85_{10}$ kg or less than $40_{10}$ kg. How much does each of them weigh? (Give your answer in base ten.)

# Chapter 16
# Computation

Any positive number can be expressed in the form $A \times 10^n$ where $1 \leqslant A < 10$ and $n$ is an integer. This is known as the standard index form of the number, e.g.

$$243\,000 = 2{\cdot}43 \times 10^5, \qquad 0{\cdot}00938 = 9{\cdot}38 \times 10^{-3}.$$

When estimating the value of an arithmetical expression, it is often useful to take each of the numbers involved in the expression to one significant figure and express each number in standard index form, e.g.

(a)
$$\frac{2710 \times 0{\cdot}0425}{0{\cdot}00532} \doteqdot \frac{3000 \times 0{\cdot}04}{0{\cdot}005}$$
$$= \frac{3 \times 10^3 \times 4 \times 10^{-2}}{5 \times 10^{-3}}$$
$$= \frac{3 \times 4}{5} \times \frac{10^3 \times 10^{-2}}{10^{-3}}$$
$$\doteqdot 2 \times 10^3 \times 10^{-2} \times 10^3 = 2 \times 10^4.$$

(b) $(0{\cdot}00713)^2 \doteqdot (7 \times 10^{-3})^2 = 49 \times 10^{-6} = 4{\cdot}9 \times 10^{-5}.$

When estimating square roots it is advisable to write the number in the form $A \times 10^n$ where $1 \leqslant A < 100$ and $n$ is even e.g.

(a) $\sqrt{235710} \doteqdot \sqrt{(24 \times 10^4)} \doteqdot 5 \times 10^2.$

(b) $\sqrt{0{\cdot}0000073} \doteqdot \sqrt{7 \times 10^{-6}} \doteqdot 3 \times 10^{-3}.$

If two numbers, given correct to two significant figures, are multiplied together then the result is not necessarily correct to two significant figures. For example, the sides of a rectangle are measured correct to two significant figures as 1·2 cm and

74

3·7 cm. If the true lengths of the sides are $x$ cm and $y$ cm, then $x$ and $y$ must satisfy the inequalities

$$1·15 < x < 1·25 \quad \text{and} \quad 3·65 < y < 3·75.$$

Thus the area of the rectangle, in $\text{cm}^2$, can lie anywhere between

$$1·15 \times 3·65 = 4·1975 \quad \text{and} \quad 1·25 \times 3·75 = 4·6875.$$

## EXERCISES

1. Express the following numbers in standard index form:

(a) 3700,  (b) 23 110,  (c) 0·035,
(d) 0·000 201,  (e) $22\frac{1}{2}$,  (f) $0·005 \times 10^6$,
(g) $(4 \times 10^{-2})^2$,  (h) $(7 \times 10^4) \times (8 \times 10^2)$,
(i) $(2 \times 10^2) \div (5 \times 10^4)$.

2. Estimate, to one significant figure, the value of

(a) $230 \times 0·00071$,  (b) $6·4/0·031$,
(c) $\sqrt{235·1}$,  (d) $\sqrt{0·00052}$,
(e) $\dfrac{2·32 \times 76·1}{316}$,  (f) $\dfrac{0·0017}{0·31 \times 0·036}$,
(g) $\sqrt{\dfrac{231\,000}{761·8}}$,  (h) $\dfrac{2·4 \times 0·045}{26·3 \times 7·8}$,
(i) $\sqrt{(3624/19)}$,  (j) $(0·2)^{10}$.

3. A manufacturer requires cartons which will hold tins of cat food. Tins are cylinders with base diameter 7 cm, correct to the nearest mm. State the length and breadth of a carton which will certainly take twelve tins arranged in three rows of four.

4. If a car travels at 50 km/h for two hours, the speed being measured to the nearest km/h and the time correct to the nearest six minutes, how accurately can the distance travelled by the car be stated?

5. A side of a square is measured as 2·2 cm correct to two significant figures. Between which two values must the area of the square lie?

If the area is obtained by squaring 2·2 cm, to how many significant figures is the answer necessarily correct?

# Chapter 17
# Logarithms

The logarithmic function $x \to \log_a x$ is the inverse of the exponential function $x \to a^x$; $a$ is the base of the logarithm.

$$a^x = y \Leftrightarrow x = \log_a y,$$

e.g. $$2^3 = 8 \Leftrightarrow 3 = \log_2 8.$$

Fig. 17.1 shows the graph of the function $x \to 2^x$. From the

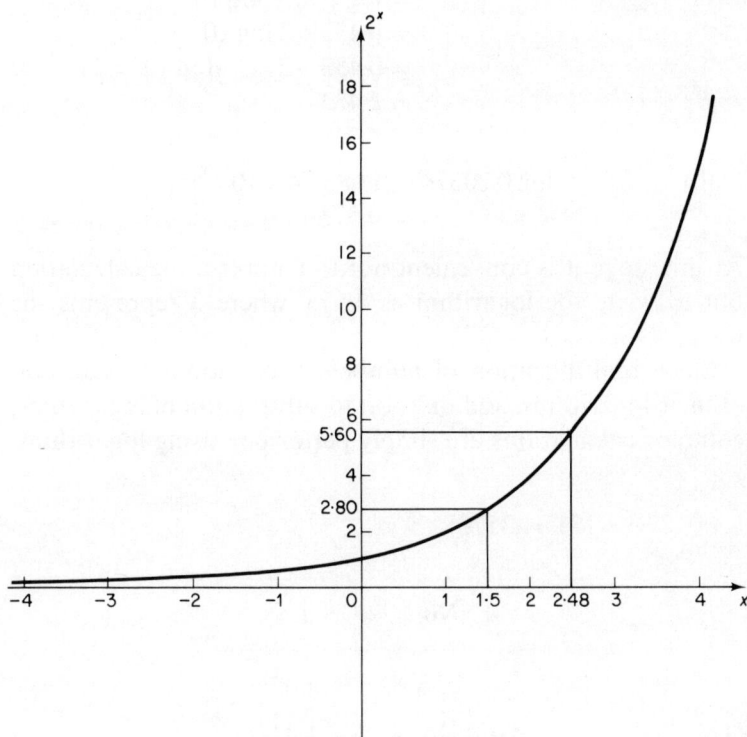

FIG. 17.1

graph $2^{1\cdot50} = 2\cdot80$, so $\log_2 2\cdot80 = 1\cdot5$. Similarly $\log_2 5\cdot6 = 2\cdot48$.

The basic properties of logarithms are

$$\text{(i)} \quad \log_a p + \log_a q = \log_a pq,$$
$$\text{(ii)} \quad \log_a p - \log_a q = \log_a (p/q),$$
$$\text{(iii)} \quad \log_a p^n = n \log_a p.$$

In computation, logarithms to the base 10 are normally used. Tables give directly the logarithm of a number between 1 and 10. For numbers outside this range it is simplest to use standard index form to find the logarithm, e.g.

$$
\begin{aligned}
\text{(a)} \qquad \log 271 &= \log(2\cdot71 \times 10^2) \\
&= \log 2\cdot71 + \log 10^2 \\
&= 0\cdot433 + 2 \log 10 \\
&= 0\cdot433 + 2 \qquad (\log 10 = 1) \\
&= 2\cdot433
\end{aligned}
$$

$$
\begin{aligned}
\text{(b)} \qquad \log 0\cdot00374 &= \log(3\cdot74 \times 10^{-3}) \\
&= 0\cdot573 + (-3)
\end{aligned}
$$

At this stage it is convenient not to complete the calculation but to write the logarithm as $\bar{3}\cdot573$, where $\bar{3}$ represents the integer $-3$.

Since multiplication of numbers corresponds to addition of their logarithms, and division to subtraction of logarithms, complex calculations are simply performed using logarithms, e.g.

(a) $27\cdot3 \times 131 = 3570$

| No. | Log. |
|:---:|:---:|
| 27·3 | 1·436 |
| 131 | 2·117 |
| $3\cdot57 \times 10^3$ | 3·553 |

(b) $236 \div 7.92 = 29.8$

| No. | Log. |
|---|---|
| 236 | 2·373 |
| 7·92 | 0·899 |
| $2.98 \times 10^1$ | 1·474 |

(c) $\dfrac{37 \cdot 6 \times 0 \cdot 0235}{9 \cdot 23} = 0 \cdot 0957$

| No. | Log. |
|---|---|
| 37·6 | 1·575 |
| 0·0235 | $\bar{2}$·371 |
| | $\bar{1}$·946 |
| 9·23 | 0·965 |
| $9.57 \times 10^{-2}$ | $\bar{2}$·981 |

## EXERCISES

1. Draw an accurate graph of the function $x \rightarrow 2^x$. Use your graph to find

      (a) $\log_2 16$,     (b) $\log_2 5 \cdot 4$,     (c) $\log_2 17 \cdot 8$.

Use your answers to calculate $5 \cdot 4 \times 17 \cdot 8$ and $17 \cdot 8 \div 5 \cdot 4$ using logarithms to the base 2.

2. Find the logarithms to the base 10 of the following numbers:

(a) 5·32,     (b) 381,     (c) 0·362,     (d) 3760,
(e) 0·00053,     (f) 98500,     (g) $7 \cdot 32 \times 10^{-6}$,     (h) $221 \times 10^4$.

3. Give numbers whose logarithms to the base 10 are

(a) 0·413,     (b) 2·716,     (c) $\bar{1}$·431,     (d) $\bar{2}$·835,
(e) 3·556,     (f) $\bar{4}$·037,     (g) 4·238,     (h) 5·000.

4. Use logarithms to calculate

(a) $36·3 \times 6·13$,

(b) $0·942 \times 27·8$,

(c) $2710 \div 13·7$,

(d) $542 \div 0·0813$,

(e) $\sqrt{38·1}$,

(f) $(0·572)^3$,

(g) $\dfrac{6290 \times 198}{372}$,

(h) $\dfrac{0·0385 \times 7·21}{356 \times 0·284}$,

(i) $\sqrt{\dfrac{213 \times 1·56}{0·0923}}$,

(j) $\sqrt[5]{2}$.

5. Find the volume of a right circular cone of base radius 2·56 cm and height 5·36 cm. What would such a cone weigh if it were made of material of density 3·51 g/cm$^3$?

6. If $t = 2\pi\sqrt{(I/mgh)}$ find the value of $t$ when $I = 134$, $m = 2·56$, $g = 9·81$ and $h = 5·21$.

7. State the value of

(a) $\log_2 8$,

(b) $\log_3 9$,

(c) $\log_4 (1/16)$,

(d) $\log_2 1024$,

(e) $\log_7 1$,

(f) $\log_3 81$,

(g) $\log_9 3$,

(h) $\log_a a^2$.

# Chapter 18
# The Slide Rule

The slide rule gives no indication of the position of the decimal point in the answer to any calculation. Hence it is essential to estimate the size of the answer using a method such as that outlined in Chapter 16.

If $x$ and $y$ are two numbers then to calculate $xy$, 1 (or 10) on the $B$ scale is set against $x$ on the $A$ scale and the answer appears on the $A$ scale opposite $y$ on the $B$ scale. Fig. 18.1 shows the setting for the calculation $2 \cdot 3 \times 1 \cdot 6$ ( $= 3 \cdot 68$).

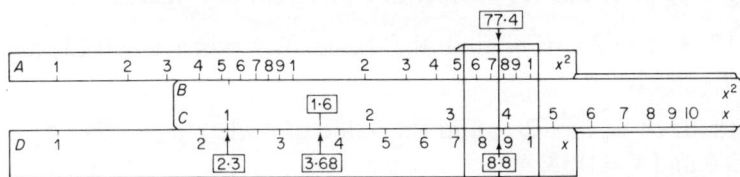

FIG. 18.1

To calculate $x \div y$, $y$ on the $B$ scale is set against $x$ on the $A$ scale and the answer appears on the $A$ scale against 1 (or 10) on the $B$ scale. Fig. 18.1 shows the setting for the calculation $3 \cdot 68 \div 1 \cdot 6$ ( $= 2 \cdot 3$).

To calculate $x^2$ the cursor is set over $x$ on the $A$ scale and the answer appears under the cursor on the $D$ scale. Fig. 18.1 shows the setting for the calculation $8 \cdot 8^2$ ( $= 77 \cdot 4$).

To calculate $\sqrt{y}$ the cursor is set over $y$ on the $D$ scale and the answer appears under the cursor on the $A$ scale. Note that as $y$ occurs twice on the $D$ scale, an estimate of the answer is vital if the correct possibility is to be chosen. Fig. 18.1 shows the setting for the calculation $\sqrt{7740}$ ( $= 88$).

81

## EXERCISES

Perform the following calculations on your slide rule.

1. $2 \cdot 32 \times 8 \cdot 51$       2. $15 \cdot 6 \div 2 \cdot 43$       3. $0 \cdot 052 \times 37$

4. $8 \cdot 13 \div 92$       5. $\sqrt{371}$       6. $\sqrt{0 \cdot 72}$

7. $\dfrac{0 \cdot 37 \times 12 \cdot 6}{45}$       8. $(0 \cdot 57)^2$       9. $7 \cdot 53 \times (2 \cdot 8)^2$

10. $\dfrac{21 \cdot 4 \times 3 \cdot 14}{0 \cdot 35 \times 105}$       11. $\left( \dfrac{0 \cdot 36 \times 76}{25 \cdot 3} \right)^2$       12. $\pi(3 \cdot 81)^2$

13. $\frac{4}{3}\pi \times (2 \cdot 38)^3$       14. $\sqrt{\dfrac{37 \cdot 6 \times 0 \cdot 58}{0 \cdot 0037}}$       15. $\dfrac{3 \cdot 76 \times (8 \cdot 1)^2}{35 \cdot 8}$

16. Find the area of a 127° sector of a circle of radius 4·62 cm, giving your answer correct to two significant figures.

17. If $t = 2\pi\sqrt{(l/g)}$ find the value of $t$ when $l = 0 \cdot 25$ and $g = 9 \cdot 81$.

18. If $v^2 = u^2 + 2fs$, find the value of $v$ when $u = 17 \cdot 4$, $f = 3 \cdot 6$ and $s = 132$.

# Chapter 19
# Formulae

*Brackets*
Expressions involving brackets can be expanded by use of the distributive law

$$a(b+c) = ab+ac$$

e.g.
$$\begin{aligned}
(3x+2)(2x-1) &= (3x+2)2x+(3x+2)(-1) \\
&= 6x^2+4x-3x-2 \\
&= 6x^2+x-2
\end{aligned}$$

This can be conveniently expressed as follows:

| $x$ | $2x$ | $-1$ |
|---|---|---|
| $3x$ | $6x^2$ | $-3x$ |
| $2$ | $4x$ | $-2$ |

$$\begin{aligned}
(3x+2)(2x-1) &= 6x^2-3x+4x-2 \\
&= 6x^2+x-2.
\end{aligned}$$

In an expression such as $2-(2x+4)(4x-3)$ it is important to notice that the whole of $(2x+4)(4x-3)$ is subtracted. Thus

$$\begin{aligned}
2-(2x+4)(4x-3) &= 2-(8x^2+10x-12) \\
&= 2-8x^2-10x+12 \\
&= -8x^2-10x+14.
\end{aligned}$$

Three important results should be remembered:

(i) $\quad (a+b)^2 = a^2+2ab+b^2,$
(ii) $\quad (a-b)^2 = a^2-2ab+b^2,$
(iii) $(a-b)(a+b) = a^2-b^2.$

*Formulae*
A formula describes the way in which the value of one variable

quantity (the subject of the formula) can be calculated from the values of other quantities, e.g. the formula

$$s = ut + \tfrac{1}{2}ft^2$$

enables the value of $s$ to be calculated when the values $u$, $t$ and $f$ are known. In this formula $s$ is the subject since $s$ can be calculated directly. If the values of $s$, $u$ and $t$ are known an equation can be obtained from which the value of $f$ can be deduced. In practice it is frequently more desirable to make $f$ the subject of the formula. The process of changing the subject of a formula is similar to that of solving equations and finding inverse functions, e.g.

$$s = ut + \tfrac{1}{2}ft^2$$

Subtracting $ut$ from each side, $\quad s - ut = \tfrac{1}{2}ft^2$.
Multiplying each side by 2, $\quad 2(s - ut) = ft^2$.
Dividing each side by $t^2$, $\quad 2(s - ut)/t^2 = f$.

## EXERCISES

1. Write without brackets

(a) $(2x+1)(3x+2)$,
(b) $(3x-2)(5x-3)$,
(c) $(4x-5)(4x+5)$,
(d) $(1-2x)(6x+2)$,
(e) $(x+2y)(2x+y)$,
(f) $(4t+7)(3t-4)$,
(g) $(3-4p)(4-2p)$,
(h) $(3a+4b)(3a+5b)$,
(i) $(3y-2)(2y^2-y+3)$,
(j) $(a+c)(b+d)$.

2. Make the given letter the subject of the formula in each of the following cases:

(a) $v = u + ft$, $\quad t$
(b) $y = ax^2 + b$, $\quad x$
(c) $v^2 = a^2(w^2 - x^2)$, $\quad x$
(d) $l = l_0(1 + at)$,
(e) $D = b^2 - 4ac$, $\quad a$
(f) $t = 2\pi\sqrt{\dfrac{l}{g}}$, $\quad l$
(g) $v^2 = w\left(\dfrac{1}{a} - \dfrac{2}{r}\right)$, $\quad r$
(h) $r = a\left[1 + \sqrt{\left(1 + \dfrac{l}{a}\right)}\right]$, $\quad l$

3. Apples cost $x$p each. Write down a formula to give the change $C$p I get from a £1 note if I buy $a$ apples. If this change is just sufficient to enable me to buy $b$ bananas at $y$p each write down another formula to give $C$ in terms of $y$ and $b$. State the relationship between $x$, $y$, $a$ and $b$.

4. If $v = u+ft$ and $s = \frac{1}{2}(v+u)t$, obtain a formula giving $v$ in terms of $u$, $f$ and $s$.

5. A square of side $2a$ has a semi-circle fixed to one of its sides so that the side of the square forms the diameter of the semi-circle. If $A$ is the area and $P$ the perimeter of the resulting figure, find expressions for (a) $a$ in terms of $P$, (b) $a$ in terms of $A$.

6. A right-angled triangle has its longest side of length $2n+1$ and one other side of length $2n-1$. Find an expression for the length of the third side and state three values of $n$ for which this expression is an integer. In general, what can you say about $n$ in this case?

7. Simplify $(x-2)^2-9$ and use your result to make $x$ the subject of the formula $y = x^2-4x-5$. Hence solve the equations

$$\text{(a) } x^2-4x-5 = 0,$$
$$\text{(b) } x^2-4x-5 = 3,$$
$$\text{(c) } x^2-4x+2 = 0.$$

Use a similar technique to solve the equations

$$\text{(d) } x^2+6x+10 = 0,$$
$$\text{(e) } x^2-2x+3 = 7.$$

# Chapter 20
# Proportionality

An ordered set is a set in which the order of elements is important. Thus $\{1, 2, 4\}$ and $\{4, 2, 1\}$ are not equal when regarded as ordered sets. If $A$ and $B$ are ordered sets, $A$ is said to be directly proportional to $B$ if there exists a function $x \rightarrow kx$ where $k$ is a constant, with domain $A$ and range $B$ which preserves the order of the sets. Thus for any $y \in B$, $y = kx$ where $x$ has the same position in set $A$ as that of $y$ in set $B$. This can also be written $y \propto x$. $k$ is called the constant of proportionality, e.g.

$$A = \{1, 2, 3, 4\} \qquad B = \{2, 4, 6, 8\}$$

Here $y \propto x$, the function $x \rightarrow 2x$ mapping $A$ onto $B$. Thus the constant of proportionality is 2 and the relation $y \propto x$ can be written $y = 2x$.

Other types of proportionality can be defined, e.g.
(a) $y$ is inversely proportional to $x$ ($y \propto 1/x$), i.e. a relation of the form $y = k/x$ connects the ordered sets,
(b) $y$ is proportional to the square of $x$ ($y \propto x^2$), i.e. a relation of the form $y = kx^2$ connects the ordered sets.

*Example*

$$A = \{2, 4, 8\} \qquad B = \{16, 8, 4\}$$

Here

$$y \propto 1/x \qquad \text{i.e. } y = k/x$$

Since

$$y = 16 \text{ when } x = 2,$$
$$16 = k/2,$$
$$k = 32.$$

The relation between the ordered sets is $y = 32/x$.

In Fig. 20.1 the results of plotting $y$ against $x$ graphically

(a)

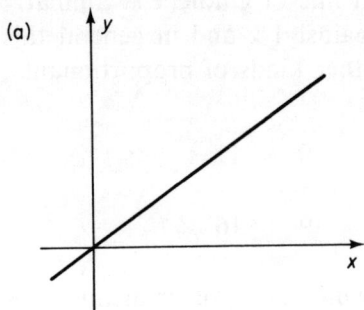

$y \propto x$
Straight line passing
through the origin, gradient $k$

(b)

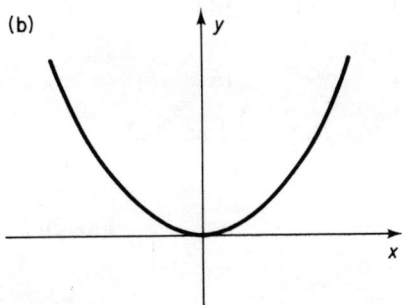

$y \propto x^2$
Parabola

(c)

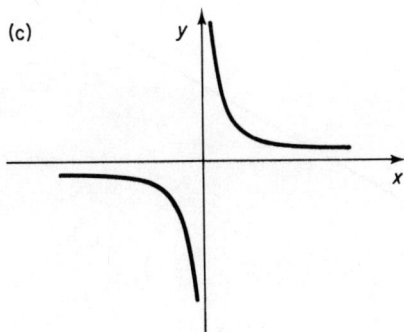

$y \propto \frac{1}{x}$
Hyperbola

FIG. 20.1

87

are shown for various types of proportion. Since in case (b) the proportionality can be written $y = kx^2$, the graph of $y$ plotted against $x^2$ is a straight line of gradient $k$. Similarly in case (c) $y$ can be plotted against $1/x$ and in general this technique can be applied to other kinds of proportionality, e.g.

| $x$ | 3 | 6 | 9 | 12 |
|---|---|---|---|---|
| $y$ | 1 | 4 | 9 | 16 |

Here $y \propto x^2$, so the table showing values of $x^2$ against $y$ is calculated.

| $x^2$ | 9 | 36 | 81 | 144 |
|---|---|---|---|---|
| $y$ | 1 | 4 | 9 | 16 |

This table yields the straight line graph in Fig. 20.2.

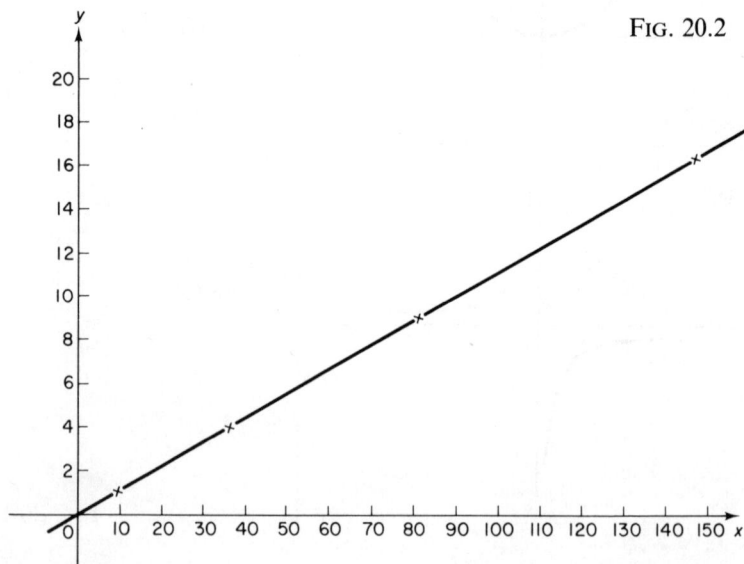

FIG. 20.2

## EXERCISES

1. Complete the following tables of values of $x$ and $y$, in each case assuming the relationship given.

(a) $y \propto x$

| $x$ | 1 | $1\frac{1}{4}$ | $3\frac{1}{2}$ | 4 | 10 |
|---|---|---|---|---|---|
| $y$ | | $2\frac{1}{2}$ | | | |

(b) $y \propto x$

| $x$ | $-5$ | $-1$ | 0 | $2\frac{1}{2}$ | $3\frac{1}{4}$ |
|---|---|---|---|---|---|
| $y$ | | | | $8\frac{3}{4}$ | |

(c) $y \propto x^2$

| $x$ | 4 | 6 | $\frac{1}{2}$ | 3 | $-1$ |
|---|---|---|---|---|---|
| $y$ | | 24 | | | |

(d) $y \propto 1/x$

| $x$ | 2 | 3 | 7 | 14 | 21 |
|---|---|---|---|---|---|
| $y$ | | | 6 | | |

(e) $y \propto 1/x^2$

| $x$ | $-10$ | $-5$ | $\frac{1}{2}$ | 2 | 3 |
|---|---|---|---|---|---|
| $y$ | | | | 25 | |

(f) $y \propto \sqrt{x}$

| $x$ | 0 | 1 | $6\frac{1}{4}$ | 8 | 9 |
|---|---|---|---|---|---|
| $y$ | | | | | 6 |

2. (a) If $y$ is proportional to $x^2$ and $y = 12$ when $x = 2$, find the value of $y$ when $x = 5$.

   (b) If $y$ is inversely proportional to $x$ and $y = 3\frac{1}{3}$ when $x = 12$, find the value of $y$ when $x = 25$.

   (c) If $y$ is proportional to $x^3$ and $y = 6$ when $x = \frac{1}{2}$, find the value of $x$ when $y = \frac{3}{4}$.

3. Investigate the type of proportionality between $x$ and $y$ in each of the following examples, and give the equation connecting $x$ and $y$:

(a)

| $x$ | 10 | 5 | 2 | 1·25 |
|---|---|---|---|---|
| $y$ | 4 | 2 | 0·8 | 0·5 |

(b)

| $x$ | 1·5 | 2·8 | 4·5 | 8·2 |
|---|---|---|---|---|
| $y$ | 1·13 | 2·12 | 3·4 | 6·2 |

(c)

| $x$ | 1·5 | 2·0 | 2·5 | 3·0 |
|---|---|---|---|---|
| $y$ | 15 | 11·25 | 9 | 7·5 |

4. The following table shows the results of an experiment.

| $x$ | 2 | 3 | 4 | 5 | 6 |
|---|---|---|---|---|---|
| $y$ | 17·2 | 26·2 | 34·6 | 44·5 | 52·2 |

Plot a graph of these results, and assuming $x$ and $y$ to be directly proportional estimate the constant of proportionality. Does this assumption seem justified?

5. In an experiment the following results are obtained.

| $x$ | 10 | 20 | 30 | 40 | 50 |
|---|---|---|---|---|---|
| $y$ | 74·0 | 294 | 642 | 1170 | 1840 |

By drawing a graph of $y$ against $x^2$ deduce the approximate relationship between $x$ and $y$.

6. The values of $x$ and $y$ shown in the table approximately satisfy the condition $y \propto 1/x^2$.

| $x$ | 0·1 | 0·2 | 0·3 | 0·4 | 0·5 |
|---|---|---|---|---|---|
| $y$ | 360 | 92 | 41 | 23 | 15 |

By drawing a suitable graph obtain the constant of proportionality.

7. The cost £$C$ of producing $x$ dinners in a certain school is approximately given by the formula $C = ax + b$, where $a$ and $b$ are constants. The table shows the total cost for various numbers of dinners.

| $x$ | 337 | 412 | 540 | 585 | 672 | 738 |
|---|---|---|---|---|---|---|
| $C$ | 53 | 58 | 68 | 73 | 78 | 84 |

Assuming that each day the total cost consists of a fixed amount to cover overheads, plus the cost of the ingredients, use a graphical method to determine

(a) the fixed charge the kitchen must pay each day,
(b) the cost of the ingredients for one meal.

Hence state the values of the constants $a$ and $b$.

8. It is known that quantities $f$, $v$ and $r$ are such that $f$ is proportional to $v$ and inversely proportional to $r$, i.e. $f$, $v$ and $r$ are connected by the formula $f = k(v^2/r)$ where $k$ is a constant. An experiment is performed in which the values of $v$ and $r$ are varied and the corresponding value of $f$ measured, to give the following results.

| $v$ | 1·1 | 1·4 | 2·3 | 3·1 | 4·7 |
|---|---|---|---|---|---|
| $r$ | 3 | 4 | 5 | 6 | 7 |
| $f$ | 1·2 | 1·5 | 3·1 | 4·8 | 9·4 |

Explain why the graph of $f$ plotted against $v^2/r$ is a straight line. Draw this straight line and find the value of $k$.

9. Variables $t$, $x$ and $a$ are connected by the formula $t = k\sqrt{(x/a)}$, where $k$ is a constant. Describe the types of proportion connecting $t$ with $x$ and $a$.

The table shows the values of $t$, $x$ and $a$ obtained in an experiment.

| $x$ | 1 | 2 | 3 | 4 | 5 |
|---|---|---|---|---|---|
| $a$ | 0·8 | 1·2 | 1·4 | 1·6 | 1·2 |
| $t$ | 6·9 | 8·1 | 9·2 | 9·9 | 12·8 |

By drawing a suitable straight line graph estimate the value of $k$.

# Chapter 21
# Calculations from Graphs

The gradient of the graph of a non-linear function at any point is the gradient of the tangent to the graph at that point (see Fig. 21.1).

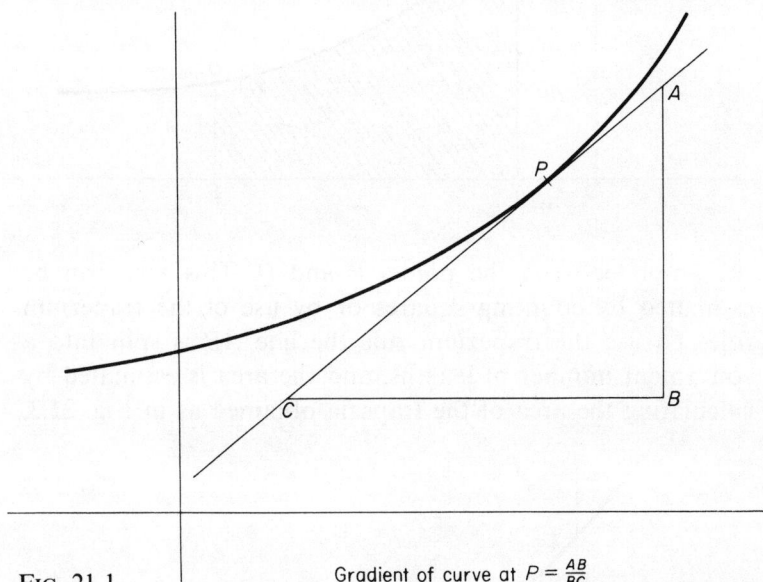

FIG. 21.1 Gradient of curve at $P = \frac{AB}{BC}$

The gradient measures the rate at which the quantity measured on the vertical axis is changing with respect to the quantity measured on the horizontal axis, e.g. if the graph in Fig. 21.1 gives the area of a forest fire (measured in km$^2$) against the time (measured in hours) the gradient gives the rate at which the area of the fire is increasing (measured in km$^2$/h).

The gradient of a distance–time graph in general gives velocity and the gradient of a velocity–time graph gives

93

acceleration (time being measured on the horizontal axis in each case).

In Fig. 21.2 the shaded area is described as the area under

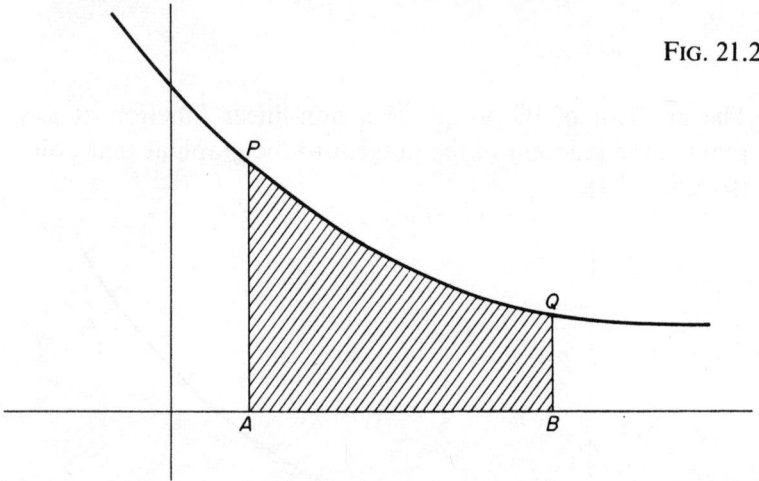

FIG. 21.2

the graph between the points $P$ and $Q$. This area can be estimated by counting squares or by use of the trapezium rule. To use the trapezium rule the line $AB$ is split into a convenient number of lengths, and the area is estimated by calculating the area of the trapezia obtained as in Fig. 21.3.

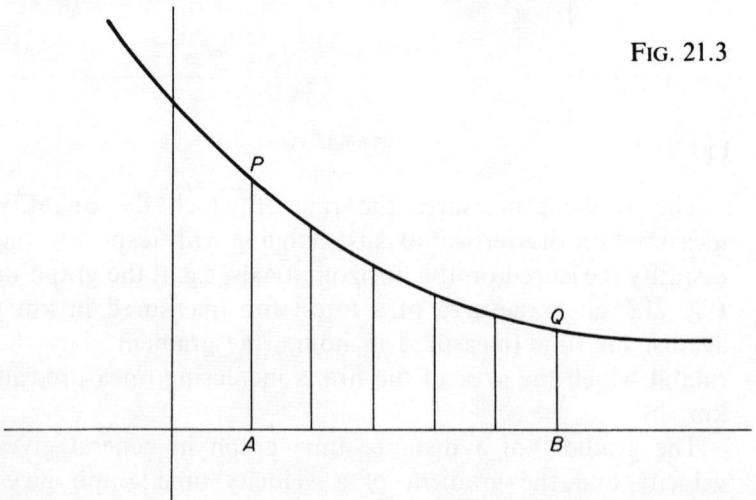

FIG. 21.3

If the vertical axis measures the rate at which a quantity $Z$ is changing with respect to the variable on the horizontal axis the area represents the increase in the quantity $Z$ during the section under consideration, e.g. if in Fig. 21.2 the vertical axis measures the rate at which the forest fire is increasing (in $km^2/h$) and the horizontal axis measures time (in hours), the area under the graph gives the increase in area of the fire between times at $A$ and $B$.

In general the area under a velocity–time graph gives distance travelled.

## EXERCISES

1. Plot the graph of $y = x^2 + 3x$ for values of $x$ between $-3$ and 4. Estimate the gradient of the graph at $x = -2$ and $x = 2$, and find the area of the region bounded by that portion of the graph between $x = -3$ and $x = 0$, and the x-axis.

2. Estimate the areas of the regions bounded by

   (a) the curve $y = 2x^2 + 6$ and the lines $x = -2$, $x = 4$ and $y = 0$,
   (b) the curve $y = 2^x$ and the lines $x = 0$, $x = 4$ and $y = 0$,
   (c) the curves $y = (x+1)^2$ and $y = 6/(x+1)$, and the lines $x = 1$ and $x = 5$,
   (d) the curve $y = x^3$ and the lines $x = 0$ and $y = 64$.

3. Draw the graph of $y = 2x + 1$, and use this graph to complete the following table showing the area under the graph between $x = 0$ and various values of $x$:

| $x$ | 1 | 2 | 3 | 4 | 5 | 6 |
|-----|---|---|---|---|---|---|
| Area | 2 | | | 20 | | |

Draw a new graph plotting these areas against $x$. Find the gradient of this graph at $x = 1, 3, 5 \ldots$ and comment on your results.

95

4. A car accelerates from rest to reach a speed of 15 m/s in 10 s. It moves steadily at this speed for a further 20 s before the driver sees a zebra crossing and begins to brake. Over the next 5 s his speed drops to 6 m/s whereupon the crossing clears and he accelerates again, reaching 12 m/s in another 5 s. He continues at this speed. Assuming that all changes of speed take place at a steady rate, draw a velocity–time graph to represent his motion over the first minute. Hence find the distance he travels in this minute.

5. The speed of a coach is noted every five minutes to give the following table:

| Time (min) | 0 | 5 | 10 | 15 | 20 | 25 | 30 |
|---|---|---|---|---|---|---|---|
| Speed (km/hour) | 65 | 69 | 79 | 77 | 63 | 54 | 48 |

Estimate the distance travelled by the coach in this half-hour, and hence find the average speed of the coach in this time.

6. The table shows the temperature of a piece of metal which has been heated and allowed to cool:

| Time (min) | 0 | 1 | 2 | 3 | 4 | 5 | 6 |
|---|---|---|---|---|---|---|---|
| Temperature (°C) | 418 | 163 | 69 | 35 | 22 | 18 | 15 |

Use a graphical method to estimate

  (a) the temperature of the metal after $1\frac{1}{2}$ minutes and $2\frac{1}{2}$ minutes,
  (b) the time at which the temperature is 200°C,
  (c) the rate at which the piece of metal is cooling after 1 minute, 4 minutes.

7. A mountain walker makes estimates of his height at various times. The results are noted as follows:

| Hours after starting | 0 | $\frac{1}{2}$ | 1 | $1\frac{1}{2}$ | 2 | $2\frac{1}{2}$ | 3 | $3\frac{1}{2}$ | 4 | $4\frac{1}{2}$ | 5 |
|---|---|---|---|---|---|---|---|---|---|---|---|
| Height (m) | 195 | 210 | 250 | 370 | 590 | 765 | 810 | 800 | 690 | 420 | 375 |

Use a graphical method to estimate:

(a) the height of the mountain, and the time he reaches the summit,

(b) the time at which he is gaining height most rapidly. (Is this necessarily the steepest part of the mountain?),

(c) the rate at which he is gaining height after 1 hour.

8. In a traffic survey hourly estimates are made of the traffic flow. The results are shown in the table:

| Time | 6·00 | 7·00 | 8·00 | 9·00 | 10·00 | 11·00 | 12·00 |
|---|---|---|---|---|---|---|---|
| Cars/hour | 20 | 180 | 510 | 570 | 440 | 290 | 260 |

Represent this information graphically, and use your graph to estimate:

(a) the maximum traffic flow,

(b) the total number of cars passing between 6.00 and 12.00.

9. The table shows the number of people in a football ground before the start of a match:

| Time | 2·00 | 2·10 | 2·20 | 2·30 | 2·40 | 2·50 | 3·00 |
|---|---|---|---|---|---|---|---|
| Number of people | 2100 | 3000 | 4200 | 6900 | 16 100 | 22 400 | 25 000 |

If all these people have entered the ground through turnstiles each of which can cope with up to 3000 people per hour, use a graphical method to determine the minimum number of turnstiles necessary.

97

10. The distance travelled by a car is shown in the table:

| Time (s) | 0 | 2 | 4 | 6 | 8 | 10 |
|---|---|---|---|---|---|---|
| Distance travelled (m) | 0 | 6·5 | 16·0 | 34·5 | 51·0 | 60·0 |

Draw a graph to show this information and use your graph to estimate:

    (a) the speed of the car after four seconds,
    (b) the average speed over the first five seconds,
    (c) the maximum speed attained by the car.

# Chapter 22
# Probability

If $S$ is the set of all possible outcomes, all equally likely, of an experiment and $A$ is the subset of outcomes satisfying a particular condition, the probability that the condition will be satisfied when the experiment is performed is $n(A)/n(S)$, and is written $P(A)$. For example, to find the probability that a number between 6 and 15 chosen at random has two digits we have

$$S = \{7, 8, 9, 10, 11, 12, 13, 14\},$$
$$A = \{10, 11, 12, 13, 14\},$$
$$P(A) = \frac{n(A)}{n(S)} = \frac{5}{8}.$$

In cases where the event under consideration can be divided into two simple events, graphical representation of the set of possible outcomes is often useful.

*Example*  A man wishes to appoint a housekeeper and a chauffeur. Of the five applicants for the housekeeper's job two are left-handed, and of the four applicants for the chauffeur's job two are left-handed. The man decides to select the successful applicants by drawing lots.

In this case the set of possible outcomes can be represented as in Fig. 22.1. The set $A$ represents the set of outcomes corresponding to both people being left-handed. Thus

$$P(A) = \frac{4}{20} = \frac{1}{5}.$$

If $A$ and $B$ are two disjoint subsets of the set of possible outcomes of an experiment, i.e. $A$ and $B$ correspond to mutually exclusive outcomes, the probability of an outcome of the experiment being a member of $A$ or $B$ is $P(A) + P(B)$. For example, in a car park containing 50 cars, 14 are blue and

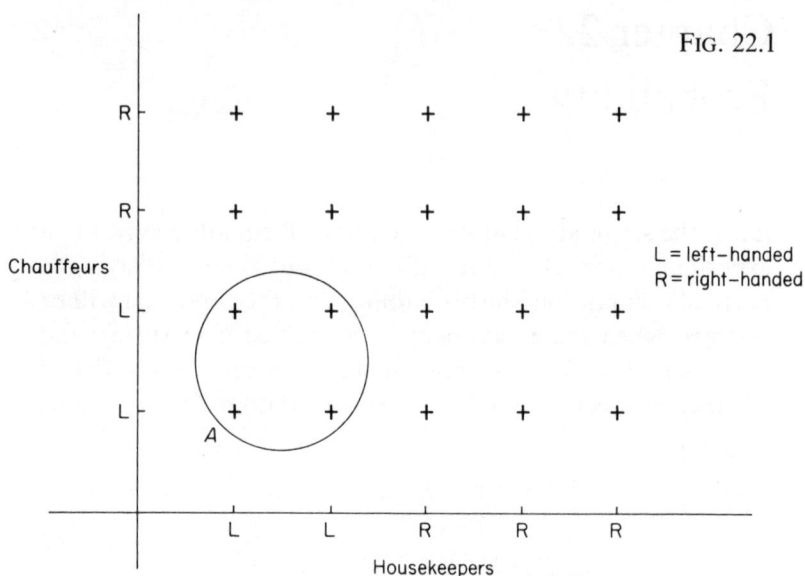

Fig. 22.1

17 are red. The probability that a car selected at random is blue or red is

$$\frac{14}{50} + \frac{17}{50} = \frac{31}{50}.$$

If two events are independent, i.e. the occurrence of one in no way affects the occurrence of the other, the probability of both events happening is the product of their separate probabilities. In the example, the probability of both applicants being left-handed is

$$\frac{1}{2} \times \frac{2}{5} = \frac{1}{5}.$$

In an experiment which consists of several different stages where the multiplication and addition laws of probability are both relevant, a tree diagram is frequently useful. A tree diagram for the above example is shown in Fig. 22.2. Note that in tracing a path along a tree diagram we multiply probabilities, and to find the probability of more than one

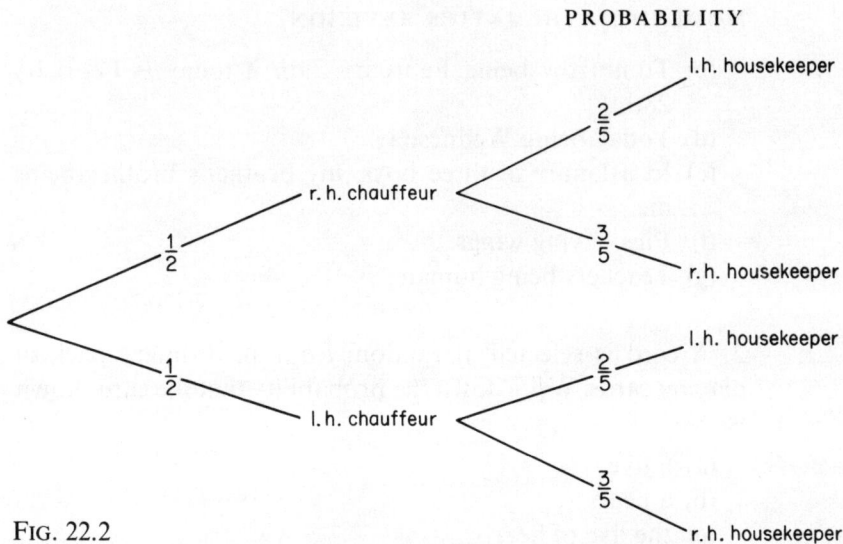

FIG. 22.2

event, as in (b), we add the corresponding probabilities. Thus an alternative solution to the example, using the tree diagram, is as follows:

Probability that both are left-handed $= \dfrac{1}{2} \times \dfrac{2}{5} = \dfrac{1}{5}.$

Probability that one is right-handed

and the other is left-handed $= \dfrac{1}{2} \times \dfrac{2}{5} + \dfrac{1}{2} \times \dfrac{3}{5} = \dfrac{1}{2}.$

Probability that both are right-handed $= \dfrac{1}{2} \times \dfrac{3}{5} = \dfrac{3}{10}.$

## EXERCISES

1. Say what you can about the probabilities of the following events.

   (a) Throwing a six on a single throw of a die.
   (b) Getting another head on a toss of a coin after nineteen successive heads have been obtained.

    (c) Tomorrow being February 29th if today is February 28th.

    (d) Today being Wednesday.

    (e) In a family of three boys, my brother's brother being me.

    (f) Pigs having wings.

    (g) Teachers being human.

2. A card is selected at random from an ordinary pack of playing cards. Write down the probability that the card drawn is

    (a) a five,

    (b) a heart,

    (c) the five of hearts,

    (d) a king, a queen or a jack,

    (e) has face value greater than four, assuming the face value of an ace is one.

3. A bag contains eight beads of which five are red and the rest blue. A bead is drawn at random from the bag. Find the probability that the bead drawn is (a) red, (b) blue.

If this bead is not replaced find the probability that the second bead drawn is a different colour to the first bead.

4. In arranging the matches for the fifth round of a knock-out competition, sixteen teams have their names placed in a hat and the names are drawn at random. If there are ten first division teams, four second division teams and two third division teams, find the probability that

    (a) a third division team will be drawn first,

    (b) the first two teams drawn will be from the first division,

    (c) if a first division team is drawn first, this team will be matched against opponents from a lower division,

    (d) there will be at least one game between first division teams,

    (e) no first division team will play at home.

5. A survey is made of the times pupils arrive at school. The result of the survey is shown in the table:

| Time | 8.20 to 8.30 | 8.30 to 8.40 | 8.40 to 8.50 | 8.50 to 9.00 | 9.00 to 9.10 | 9.10 to 9.20 |
|---|---|---|---|---|---|---|
| Number of pupils arriving in this time | 24 | 56 | 171 | 232 | 12 | 5 |

School starts at 9.00. Find the probability that a pupil chosen at random

(a) arrives at school before 8.30,

(b) is late for school,

(c) arrives at school between 8.30 and 9.00.

6. The table shows the composition of 150 families:

*No. of boys*

|  |  | 0 | 1 | 2 | 3 | 4 |
|---|---|---|---|---|---|---|
| *No. of girls* | 0 | 36 | 4 | 5 | 6 | 2 |
| | 1 | 5 | 11 | 19 | 9 | 1 |
| | 2 | 4 | 17 | 12 | 0 | 0 |
| | 3 | 7 | 7 | 1 | 0 | 0 |
| | 4 | 1 | 2 | 0 | 1 | 0 |

Thus, for example, there are nineteen families consisting of two boys and one girl. Find the probability that in a family chosen at random, there are

(a) two girls and one boy,

(b) an equal number of boys and girls (assuming the family has children),

(c) no boys.

103

7. An unbiased die has faces numbered 0 to 5. It is thrown twice and the two digits so obtained are used in the order thrown to form a number to base six. Find the probability that the number

   (a) is equal to $36_{10}$,
   (b) consists of two equal digits,
   (c) is less than $15_{10}$,
   (d) is prime.

8. A town occupies an area which can be considered to be a disc of diameter 3 km. A bomb which destroys everything within 2 km of the place where it lands is dropped at random, landing within the town boundaries. Find the probability that the whole town is destroyed.

9. When playing darts I reckon that on my first throw I can score 20 with probability $\frac{3}{5}$. If I score 20 with one throw this increases my confidence and the probability that I score 20 with my next throw becomes $\frac{3}{4}$, whereas if I miss, the probability that I score 20 with my next throw becomes $\frac{1}{2}$. Assuming that I aim for 20 each time, find the probability that in three throws I score (a) three twenties, (b) only one twenty.

10. In a cricket match Jim, the last man of one side, comes to the wicket with his team needing seven runs to win and two balls of the match left. Being renowned for his idleness he refuses to run and only scores by hitting boundaries. On receiving a ball the probability that

   (a) he is out is $\frac{1}{2}$,
   (b) he fails to score but is not out is $\frac{1}{4}$,
   (c) he hits a four is $\frac{1}{6}$,
   (d) he hits a six is $\frac{1}{12}$.

Find the probability that

   (a) Jim's team wins,
   (b) Jim's team loses,

(c) the match is tied (i.e. Jim is out when the scores are level),

(d) the match is drawn (i.e. Jim is not out but his team has not overtaken the opponents' score).

11. From past experience I know that my refrigerator will keep cheese fresh for seven days with probability $\frac{1}{2}$, and for eight days with probability $\frac{1}{10}$. I have some cheese in the refrigerator that is fresh after seven days. What is the probability that it will still be fresh after one more day?

# Chapter 23
# Statistics

The main methods of representing information diagrammatically are:

(a) *The histogram (bar chart)*
The bar chart is a series of bars or columns, the height of each bar being proportional to the frequency of occurrence of a particular event. A histogram is a similar type of representation where the area of each column is proportional to the frequency.

(b) *The pie chart*
Here the frequency of occurrence of a particular event is represented by a sector of a circle, the angle of the sector being proportional to the frequency of the event.

(c) *The pictogram*
Here the frequency of a particular event is represented by a number (including fractions) of appropriate symbols.

(d) *The line-graph*
A line-graph is obtained by plotting the information given using suitable axes and connecting each point to its neighbouring points by straight lines. Information plotted in this way normally involves time and the graph may be plotted as a broken line to indicate a general trend, or as a continuous line if points between the ones originally plotted can have any meaning.

To describe a particular population (i.e. the set of numbers under consideration) measures of position, or averages, are generally given. Such measures are:

(a) *The median*

This is the 'central' measure in that 50% of the population lies above the median and 50% lies below, i.e. for a population with $n$ members the median is the $\frac{1}{2}(n+1)$th member of the population when the population is arranged in order of size. In some cases, especially if the information is given in the form of a grouped frequency table, it is advisable to draw a cumulative frequency curve (ogive) to represent the information and to estimate the median from this curve.

*Example*  The table shows the marks obtained by 100 pupils in a certain examination and a third column has been included to give the cumulative frequency (i.e. the number of pupils scoring less than the relevant mark).

| Mark | No. of pupils | Cumulative frequency |
|------|---------------|----------------------|
| 1–10 | 12 | 12 |
| 11–20 | 14 | 26 |
| 21–30 | 42 | 68 |
| 31–40 | 22 | 90 |
| 41–50 | 10 | 100 |

This information is shown in the ogive in Fig. 23.1. Note that the values of the cumulative frequency are plotted at the ends of the relevant class intervals. In this case the intervals are $\frac{1}{2}$–$10\frac{1}{2}$, $10\frac{1}{2}$–$20\frac{1}{2}$, etc. The estimate of the median is obtained by reading from the ogive the mark corresponding to the $50\frac{1}{2}$th person, since the $50\frac{1}{2}$th person (if he existed) would divide the total of 100 people into two equal halves.

The median in this case is 27.

The quartiles are the $\frac{1}{4}$ and $\frac{3}{4}$ marks of the population and these are also shown in Fig. 23.1. The first quartile is 21 and the third quartile is 33.

The difference between the quartiles (the interquartile range) is often used as a measure of the dispersion or spread

107

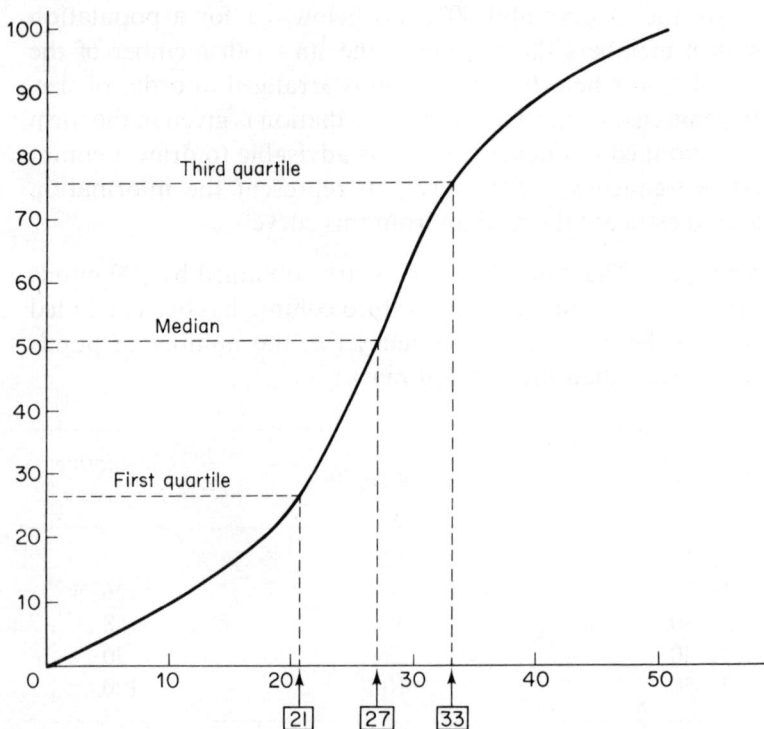

FIG. 23.1

of the population. In this case the interquartile range is $33 - 21$
$= 12$.

(b) *The mean*

This is commonly termed 'the average' and is obtained by
dividing the sum of all the members of the population by the
number of members of the population. To help avoid unneces-
sary computation it is often simpler to choose an arbitrary
origin and to perform the calculation using such an origin.

*Example* The table shows the number of shots taken by a
golfer to complete an eighteen-hole round, at various times.

108

| No. of shots taken | Frequency f | Shots taken origin = 73 x | fx |
|---|---|---|---|
| 70 | 1 | −3 | −3 |
| 71 | 3 | −2 | −6 |
| 72 | 8 | −1 | −8 |
| 73 | 5 | 0 | 0 |
| 75 | 3 | 2 | 6 |
| 78 | 2 | 5 | 10 |
| 84 | 1 | 11 | 11 |
| 86 | 1 | 13 | 13 |
| | 24 | | 40 − 17 = 23 |

Thus $$\text{mean} = 73 + \frac{23}{24} \text{ shots} = 73 \cdot 96 \text{ shots.}$$

The same answer should always be obtained regardless of the choice of the origin.

In cases where the information is given in class intervals an estimate of the mean can be obtained by considering all the members of a particular interval to have a value equal to the middle value of the interval.

(c) *The mode*
This is the 'most popular' member of the population, i.e. that member which has the greatest frequency, e.g. in the example concerning the golfer the mode of the population is 72 shots.

## EXERCISES

1. In a darts game a particular player's scores are as follows:

60, 82, 21, 79, 34, 36, 43, 43, 101, 5, 30.

Calculate
  (a) the median score,       (b) the mean score,
  (c) the mode score.

2. On a journey a driver kept a check on the distance he travelled on various types of roads and his results were as follows:

| Type of road | km travelled |
| --- | --- |
| 2-laned 'A' road | 80 |
| 3-laned 'A' road | 17 |
| 'B' road | 33 |
| Dual carriageway | 18 |
| Motorway | 42 |

Represent this information using
  (a) a pie chart,     (b) a bar chart,     (c) a pictogram.

3. For a population consisting of all the integers from 90 to 112 inclusive, state

  (a) the mean,     (b) the median,     (c) the mode,
  (d) the interquartile range.

4. In a survey the annual salaries of a group of men were noted as follows:

| Salary (£) | No. of men | Salary (£) | No. of men |
| --- | --- | --- | --- |
| 600– 800 | 4 | 1800–2000 | 42 |
| 800–1000 | 9 | 2000–2200 | 39 |
| 1000–1200 | 14 | 2200–2400 | 23 |
| 1200–1400 | 38 | 2400–2600 | 12 |
| 1400–1600 | 61 | 2600–2800 | 10 |
| 1600–1800 | 60 | 2800–4000 | 8 |

Draw the cumulative frequency curve for this information and use your graph to estimate

(a) the median salary,
(b) the interquartile range.

5. Over the first half of a season a boy keeps a check on the times he records when running 100 m and the results are shown in the table.

| Time (s) | 11·8 | 11·9 | 12·0 | 12·1 | 12·2 | 12·3 | 12·4 | 12·5 |
|---|---|---|---|---|---|---|---|---|
| No. of runs | 2 | 5 | 7 | 12 | 3 | 5 | 4 | 2 |

Calculate his mean time for the 100 m during this part of the season.

The times for his next 8 runs were

$$12·1, \ 12·1, \ 12·3, \ 12·2, \ 12·0, \ 11·8, \ 12·4, \ 11·9.$$

Calculate the mean time for these runs. Is the boy justified in claiming he has improved his performance?

6. A group of 100 pupils were asked to estimate the distance between two points and the results were as shown in the table.

| Distance (m) | 570–580 | 580–590 | 590–600 | 600–610 | 610–620 | 620–630 | 630–640 |
|---|---|---|---|---|---|---|---|
| No. of pupils | 2 | 9 | 18 | 42 | 15 | 10 | 4 |

Estimate the mean distance and explain why your answer is only an estimate.

7. During a school year a teacher gives his class one test per term and an end of year examination. The marks of three boys were as follows (each mark is given as a percentage):

| | Test 1 | Test 2 | Test 3 | Exam |
|---|---|---|---|---|
| Alan | 75 | 52 | 63 | 50 |
| Brian | 54 | 46 | 65 | 75 |
| Charles | 81 | 61 | 68 | 40 |

Calculate the mean score for each boy.

If it was decided to make the mark for the end of year examination three times as important as the mark for any individual test, calculate a new mean mark for each boy which reflects this decision.

8. The marks obtained by 200 pupils in their end of term examination are given in the table.

| Mark (%) | 20 and less | 21–30 | 31–40 | 41–50 | 51–60 | 61–70 | 71–80 | above 80 |
|---|---|---|---|---|---|---|---|---|
| No. of pupils | 5 | 10 | 19 | 30 | 47 | 54 | 28 | 7 |

It is decided to issue the marks as grades A, B, C, D and E rather than percentages. The top 10% of the pupils are given grade A, the next 20% grade B, the next 40% grade C, the next 20% grade D and the rest grade E. Draw a cumulative frequency to estimate the range of marks corresponding to each grade.

# Chapter 24
# Area and Volume

The table shows how to obtain the areas of some common figures:

| Figure | Area ($A$) | |
| --- | --- | --- |
| rectangle | length × breadth | $A = ab$ |
| parallelo-gram | base × perpen-dicular height | $A = bh$ |
| triangle | $\frac{1}{2}$base × height | $A = \frac{1}{2}bh$ <br> $= \frac{1}{2}ab \sin C$ |
| trapezium | $\frac{1}{2}$ sum of parallel sides × height | $A = \frac{1}{2}h(a + b)$ |
| circle | $\pi$ × radius squared | $A = \pi r^2$ |

The area of the parallelogram is obtained by shearing the rectangle, of the triangle by dividing the parallelogram into two congruent triangles, and of the trapezium by dividing it into two non-congruent triangles.

The surface area of a prism (i.e. a solid of constant cross

section) excluding the ends, is the perimeter of the cross section multiplied by the height, e.g. the area of the curved surface of a cylinder of base radius $r$ and height $h$ is $2\pi rh$, and the total surface area of this cylinder is $2\pi rh + 2\pi r^2$.

The surface area of a sphere is $4\pi r^2$.

| Solid | Volume (V) | |
|-------|-----------|---|
| cuboid | length × breadth × height | $V = abc$ |
| prism | area of base × height | e.g. $V = \pi r^2 h$ |
| pyramid | $\frac{1}{3}$ area of base × height | e.g. $V = \frac{1}{3}\pi r^2 h$ $V = \frac{1}{3}abh$ |
| sphere | $\frac{4}{3}\pi$ × radius cubed | $V = \frac{4}{3}\pi r^3$ |

## EXERCISES

1. A man has a rectangular garden 10 m wide by 30 m long. He decides to leave a border of width 1 m round the edge of the garden and to build a circular pool of radius 2 m. The rest he makes into a lawn. Find the area of the lawn.

114

If the depth of the pool is 75 cm, find the volume of water required to fill the pool.

2. In painting a rectangular wall 2·4 m high and 4·7 m long, I use 120 cm³ of paint. Estimate the thickness of the paint on the wall.

3. Fig. 24.1 shows the cross section of a girder of length 11 m. Find the volume of metal in the girder. If the density of the metal is 5·6 g/cm³, find the weight of the girder.

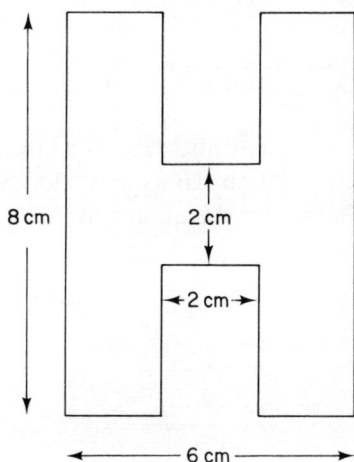

Fig. 24.1

4. A cylindrical vessel of base radius $5\frac{1}{3}$ cm contains water. When a sphere is completely immersed in the water, the water level rises by 3 cm. Find the radius of the sphere.

5. A quadrilateral has vertices $P(2, 3)$, $Q(4, 11)$, $R(11, 9)$ and $S(7, 2)$. If $A$, $B$, $C$ and $D$ are the points $(2, 0)$, $(4, 0)$, $(7, 0)$ and $(11, 0)$ respectively, find the area of the quadrilateral $PQRS$ by considering the areas of the trapezia $APQB$, $BQRD$, $APSC$ and $CSRD$.

6. Using a technique similar to that of question 5, find the areas of the following figures.

(a) the triangle with vertices $(3, 2)$, $(5, 9)$ and $(8, 7)$,

115

  (b) the quadrilateral with vertices $(0, 3)$, $(3, 8)$, $(7, 7)$ and $(9, 1)$,

  (c) the quadrilateral with vertices $(0, 0)$, $(5, 10)$, $(-2, 6)$ and $(-6, 1)$,

  (d) the quadrilateral with vertices $(4, 5)$, $(5, -6)$, $(-3, -5)$ and $(-2, 1)$.

7. A right circular cone is divided into three portions of equal height by twice cutting it parallel to the base. Determine the ratio of the volume of each portion to the volume of the cone.

8. $ABCD$ is a trapezium with $AB$ parallel to $DC$. The diagonals $BD$ and $AC$ intersect at $X$. Explain why the triangles $AXD$ and $BXC$ are equal in area.

9. Fig. 24.2 shows a quadrant $ABC$ of a circle, centre $B$, and a semi-circle constructed on $AB$ as diameter. Show that the arc of the semi-circle divides the quadrant into two equal areas.

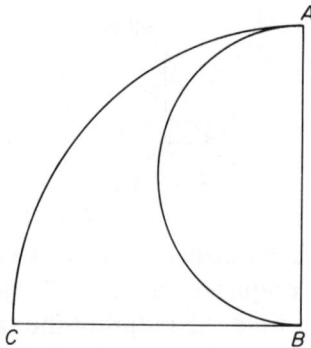

Fig. 24.2

# Chapter 25
# The Circle and the Sphere

A circle is the set of all points in a plane equidistant from a given point, called the centre of the circle. In Fig. 25.1 $O$ is the

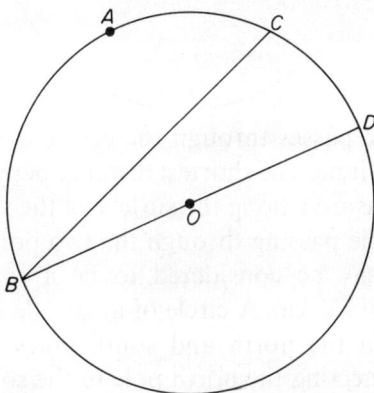

FIG. 25.1

centre of the circle, $OA$, $OB$, $OC$ and $OD$ are radii, the straight line $BC$ is a chord, the curved line $CAB$ is an arc and the straight line $BD$ is a diameter. The area bounded by the chord $BC$ and the arc $BAC$ is a segment. The area bounded by the radii $OB$ and $OC$ and the arc $BAC$ is a sector.

For a circle of radius $r$,

$$\text{Length of perimeter (circumference)} = 2\pi r,$$
$$\text{Area bounded by circle} = \pi r^2.$$

In Fig. 25.2,

$$\text{Length of arc } AXB = \frac{\theta}{360} \times 2\pi r,$$
$$\text{Area of sector } OAXB = \frac{\theta}{360} \times \pi r^2.$$

A sphere is the set of all points in space equidistant from a fixed point. The intersection of a plane and a sphere is a circle,

117

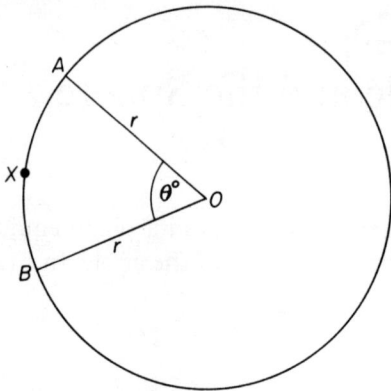

FIG. 25.2

and if the plane passes through the centre of the sphere, the circle is a great circle. The shortest distance between two points of a sphere, measured along the surface of the sphere, is an arc of the great circle passing through the two points.

The earth may be considered to be a sphere of radius approximately 6370 km. A circle of longitude is a great circle passing through the north and south poles. A meridian is semi-circle connecting the north pole to the south pole.

In specifying the location of a meridian, the meridian passing through Greenwich is taken as the base. The longitude of a point on a meridian is the angle between the plane of the meridian and the plane of the Greenwich meridian, measured east or west of Greenwich. In Fig. 25·3, the longitude of $P$ is $\theta°$ E.

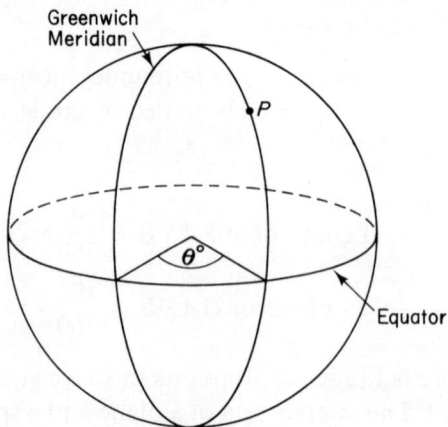

FIG. 25.3

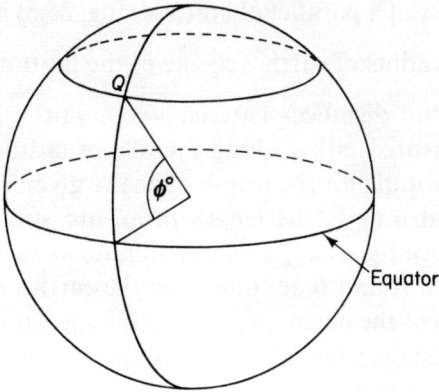

FIG. 25.4

A parallel of latitude is a circle on the surface of the earth in a plane parallel to the equator. The latitude of a point on a parallel is the angle between the radius of the earth passing through that point and the plane of the equator, or the plane of the parallel (measured north or south of the equator). In Fig. 25.4, the latitude of $Q$ is $\phi° N$.

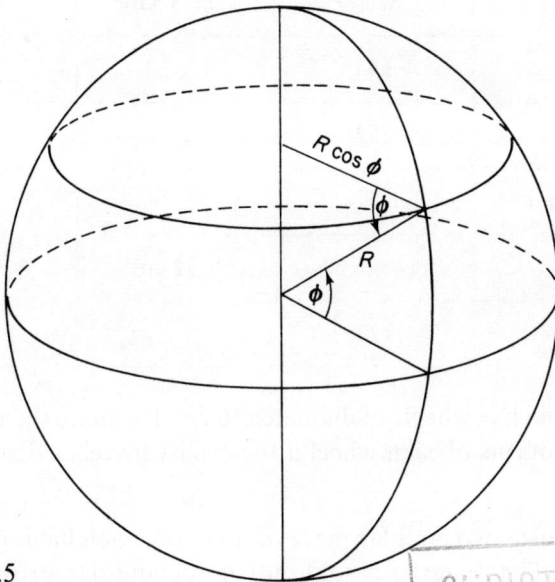

FIG. 25.5

The radius of a parallel of latitude (Fig. 25.5) is given by

radius of earth × cosine of the latitude.

To find the distance between two points on the earth's surface measured either along a circle of latitude or a great circle, the formula for the length of an arc given earlier is used.

A nautical mile is the length of an arc subtended at the earth's surface by an angle of one minute at the centre of the earth. Thus if $A$ and $B$ are points on the earth's surface and $O$ is the centre of the earth, the angle $A\hat{O}B$ measured in minutes gives the distance between the points $A$ and $B$ in nautical miles.

## EXERCISES

1. Complete the following table:

| Radius of Circle | Angle of Sector | Perimeter of Sector | Area of Sector |
|---|---|---|---|
| 5 cm | 144° | | |
| 8 cm | 252° | | |
| 2·51 cm | 77° | | |
| 6 cm | | 22 cm | |
| | 120° | | 100 cm² |

2. A cycle has wheels of diameter 56 cm. Estimate the number of revolutions of each wheel if the cyclist travels a distance of 1 km.

3. I require two circular pieces of material each having radius 600 cm. The material is sold only in rectangular strips which

120

can be bought in any size. What are the dimensions of the smallest strip I can buy and how much material will be wasted?

4. Find the shortest distance measured along the earth's surface between the following pairs of towns:

(a) Addis Ababa, Ethiopia (latitude 9° 2′ N, longitude 38° 46′ E) and Asmara, Ethiopia (latitude 15° 20′ N, longitude 38° 46′ E),

(b) Yokohama, Japan (latitude 35° 25′ N, longitude 139° 32′ E) and Bentinck Islands, Australia (latitude 17° 15′ S, longitude 139° 32′ E),

(c) Lima, Peru (latitude 12° 10′ S, longitude 77° 3′ W) and Washington, U.S.A. (latitude 38° 58′ N, longitude 77° 3′ W).

5. Find the shortest distance measured along a parallel of latitude between the following pairs of towns:

(a) Mecca, Saudi Arabia (latitude 21° 18′ N, longitude 40° 13′ E) and Raipur, India (latitude 21° 18′ N, longitude 81° 42′ E),

(b) Buenos Aires, Argentine (latitude 34° 52′ S, longitude 58° 37′ W) and Adelaide, Australia (latitude 34° 52′ S, longitude 138° 43′ E),

(c) Edmonton, Canada (latitude 53° 40′ N, longitude 113° 30′ W) and Wakefield, England (latitude 53° 40′ N, longitude 1° 30′ W).

6. Find the shortest distance along the earth's surface between the following pairs of towns:

(a) Wellington, New Zealand (latitude 41° 15′ S, longitude 176° 46′ E) and Sidmouth, England (latitude 50° 41′ N, longitude 3° 14′ W),

(b) Calcutta, India (latitude 22° 38′ N, longitude 88° 21′ E) and Cedar Rapids, U.S.A. (latitude 42° N, longitude 91° 39′ W).

7. A boat sails for 480 nautical miles due east from a position

whose longitude is 41° W and whose latitude is 29° S. Give the new position of the boat.

The boat now sails 800 nautical miles due south. Give the position of the boat after this stage of its journey.

8. Fig. 25.6 shows the target for a shooting competition. The target consists of three concentric circles, the outer circle being of diameter 30 cm. Calculate the radii of the other two circles, assuming that the shaded regions are equal in area.

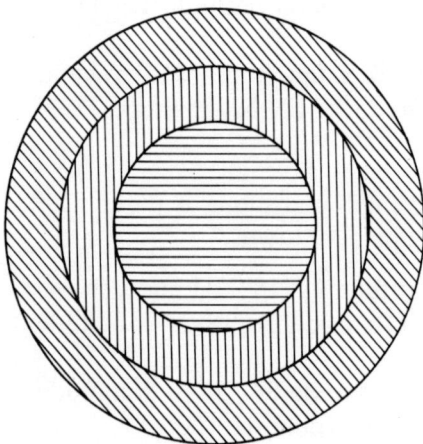

FIG. 25.6

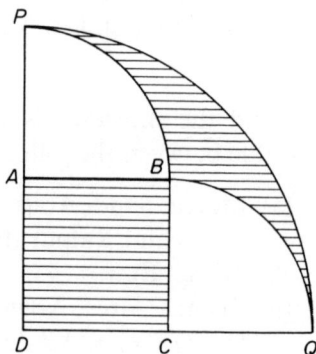

FIG. 25.7

9. In Fig. 25.7 *ABCD* is a square, *PAB*, *BCQ* and *PDQ* are quadrants of circles. Show that the length of the arc *PQ* is equal to the sum of the lengths of the arcs *PB* and *BQ*.

Show that the sum of the areas of the shaded regions is equal to half the area of the quadrant *PDQ*.

10. A satellite is put into an orbit directly over the equator and makes one complete revolution (i.e. traces out a complete circle) in ninety minutes. The satellite is observed to be directly above a point whose longitude is 20° E. If the satellite is rotating in the same direction as the earth, state the longitude of the point on the earth's surface directly below the satellite ninety minutes later.

# Chapter 26
# Topology

A topological transformation has the properties that

  (a) a line is mapped onto a line,
  (b) the order of points on a line is invariant,
  (c) the number of half-lines meeting at a point is invariant

under the transformation.

In general a topological transformation preserves the way in which points of a figure are connected to each other, but not the shape or size of the figure.

Two figures are topologically equivalent if one figure can be topologically transformed into the other. Any shape topologically equivalent to a circle is a simple closed curve.

In Fig. 26.1 (a) and (e) are topologically equivalent, (b) and (d) are simple closed curves.

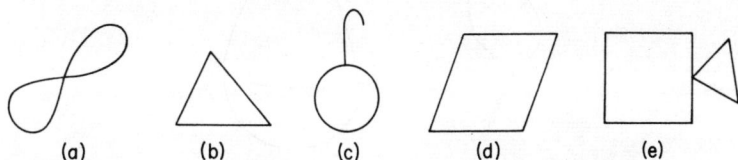

|  (a)  |  (b)  |  (c)  |  (d)  |  (e)  |

FIG. 26.1

In any figure a point where lines meet is called a node, a line joining two nodes is called an arc, and an area bounded by arcs is called a region. In Fig. 26.2, $A$, $B$, $C$, $D$ are nodes; 1, 2, 3, 4, 5 are arcs; $x$, $y$, $z$ are regions.

Fig. 26.2 is an example of a network. In a network it is possible to trace a path from any node to any other node *via* an arc or series of arcs. If a figure has $N$ nodes, $A$ arcs and $R$ regions then

$$R + N = A + 2.$$

The order of a node is the number of half-lines meeting at that node, e.g. in Fig. 26.2 *A* is a 3-node, *D* is a 1-node.

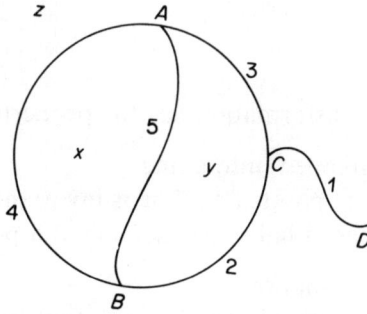

FIG. 26.2

A network may be described by means of a matrix whose elements give the number of ways of tracing a route from one node to another *via* a single arc, e.g. for the network in Fig. 26.3 we obtain the matrix

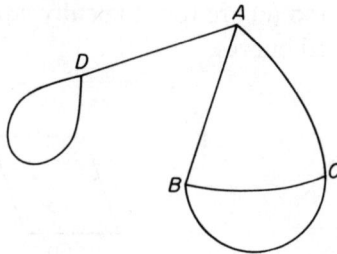

FIG. 26.3

$$\begin{array}{c@{\qquad}cccc} & A & B & C & D \\ A & \begin{pmatrix} 0 & 1 & 1 & 1 \\ B & 1 & 0 & 2 & 0 \\ C & 1 & 2 & 0 & 0 \\ D & 1 & 0 & 0 & 2 \end{pmatrix} \end{array}$$

Note that there are two routes from *D* to *D* as the loop can be traversed in either direction.

An incidence matrix describes the network in terms of any two of nodes, arcs and regions. In Fig. 26.2 arc 5 is incident on regions $x$ and $y$, node $C$ is incident on arcs, 1, 2 and 3, and region $y$ is incident on nodes $A$, $B$ and $C$, etc. In general the incidence matrix has entries 0 or 1, corresponding to non-incidence or incidence respectively. Where an arc is in the form of a loop, it is incident twice on the relevant node and the corresponding entry in the matrix is 2, e.g. for Fig. 26.2 we have

(a)

$$
\begin{array}{c}
\text{nodes}
\end{array}
\quad
\begin{array}{cc}
 & \text{arcs} \\
 & \begin{array}{ccccc} 1 & 2 & 3 & 4 & 5 \end{array} \\
\begin{array}{c} A \\ B \\ C \\ D \end{array}
&
\begin{pmatrix}
0 & 0 & 1 & 1 & 1 \\
0 & 1 & 0 & 1 & 1 \\
1 & 1 & 1 & 0 & 0 \\
1 & 0 & 0 & 0 & 0
\end{pmatrix}
\end{array}
$$

(b)

$$
\begin{array}{c}
\text{arcs}
\end{array}
\quad
\begin{array}{cc}
 & \text{regions} \\
 & \begin{array}{ccc} x & y & z \end{array} \\
\begin{array}{c} 1 \\ 2 \\ 3 \\ 4 \\ 5 \end{array}
&
\begin{pmatrix}
0 & 0 & 1 \\
0 & 1 & 1 \\
0 & 1 & 1 \\
1 & 0 & 1 \\
1 & 1 & 0
\end{pmatrix}
\end{array}
$$

(c)

$$
\begin{array}{c}
\text{regions}
\end{array}
\quad
\begin{array}{cc}
 & \text{nodes} \\
 & \begin{array}{cccc} A & B & C & D \end{array} \\
\begin{array}{c} x \\ y \\ z \end{array}
&
\begin{pmatrix}
1 & 1 & 0 & 0 \\
1 & 1 & 1 & 0 \\
1 & 1 & 1 & 1
\end{pmatrix}
\end{array}
$$

125

The transpose of the nodes-arcs incidence matrix gives the arcs-nodes incidence matrix etc.

If two networks have the same incidence matrix they are topologically equivalent.

# EXERCISES

1. State which of the shapes in Fig. 26.4 are topologically equivalent to each other.

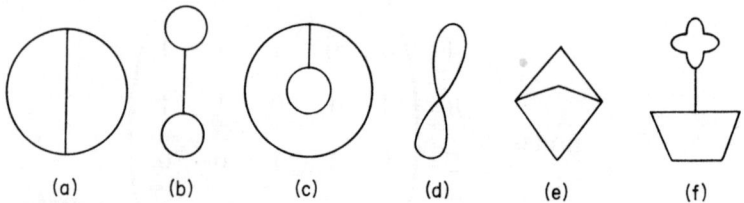

|       |       |       |       |       |       |
|-------|-------|-------|-------|-------|-------|
| (a)   | (b)   | (c)   | (d)   | (e)   | (f)   |

FIG. 26.4

2. Draw, where possible, a plane network with

    (a) one 5-node, one 4-node and two 3-nodes,
    (b) five arcs and two regions,
    (c) one 4-node, two 3-nodes and two 1-nodes,
    (d) one arc, two nodes and one region,
    (e) two regions, two arcs and three nodes.

3. Compile direct route matrices for each of the networks in Fig. 26.5.

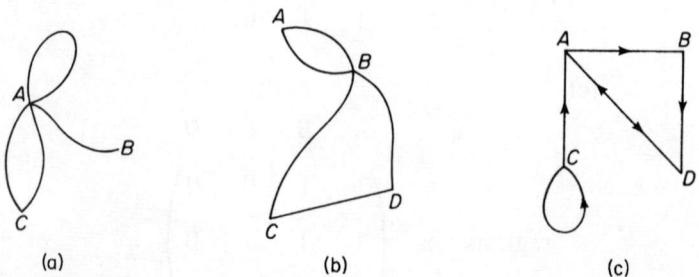

|       |       |       |
|-------|-------|-------|
| (a)   | (b)   | (c)   |

FIG. 26.5

4. How can the direct route matrix of a directed network be recognized? Construct the networks whose direct route matrices are

(a) $\begin{pmatrix} 0 & 1 & 1 & 2 & 0 \\ 1 & 0 & 1 & 0 & 1 \\ 1 & 1 & 0 & 0 & 1 \\ 2 & 0 & 0 & 0 & 0 \\ 0 & 1 & 1 & 0 & 2 \end{pmatrix}$     (b) $\begin{pmatrix} 0 & 1 & 0 & 1 \\ 1 & 2 & 1 & 0 \\ 0 & 0 & 0 & 1 \\ 1 & 1 & 1 & 0 \end{pmatrix}$

5. Draw networks represented by the following incidence matrices:

(a)

arcs $\begin{matrix} & \text{nodes} \\ \begin{pmatrix} 1 & 1 & 0 \\ 0 & 1 & 1 \\ 1 & 0 & 1 \\ 2 & 0 & 0 \\ 0 & 2 & 0 \end{pmatrix} \end{matrix}$

(b)

nodes $\begin{matrix} & \text{regions} \\ \begin{pmatrix} 1 & 1 & 1 & 1 \\ 1 & 0 & 0 & 1 \\ 1 & 1 & 1 & 1 \\ 1 & 1 & 0 & 0 \end{pmatrix} \end{matrix}$

(c)

regions $\begin{matrix} & \text{arcs} \\ \begin{pmatrix} 1 & 1 & 1 & 0 & 0 & 0 \\ 0 & 1 & 0 & 1 & 0 & 1 \\ 1 & 0 & 0 & 0 & 1 & 1 \\ 0 & 0 & 1 & 1 & 1 & 0 \end{pmatrix} \end{matrix}$

6. For the network in Fig. 26.6 construct

  (a) the nodes-arcs incidence matrix **X**,
  (b) the arcs-regions incidence matrix **Y**,
  (c) the regions-nodes incidence matrix **Z**,
  (d) the direct route matrix **M**.

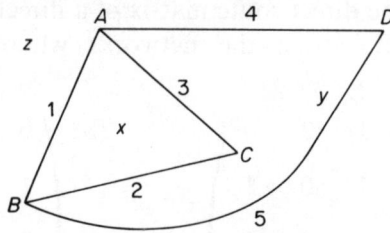

FIG. 26.6

Calculate **XX′** and **XY**. State the connection between
(i) **XX′** and **M**,     (ii) **XY** and **Z′**.

7. Fig. 26.7 shows the main roads connecting four towns.

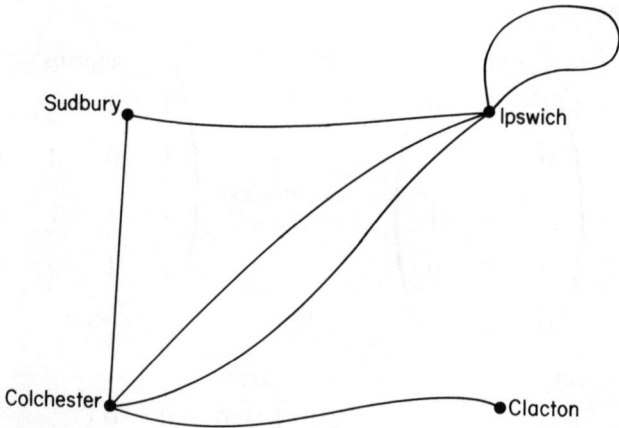

FIG. 26.7

Construct a direct route matrix **R** for this network and calculate **R²**. How many ways are there of getting from Sudbury
to Colchester, passing through just one of the towns shown?
Where does this number occur in the matrix **R²**? Deduce the
number of two stage routes

   (a) from Colchester to Ipswich,
   (b) starting and finishing at Ipswich,
   (c) starting at Sudbury.

8. Fig. 26.8 shows the plan of a maisonette. Is it possible to

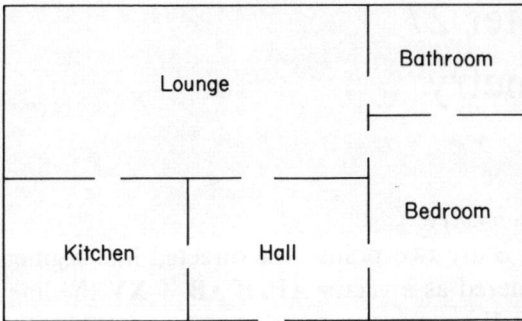

FIG. 26.8

make a tour of the maisonette using each door only once, starting

(a) from the kitchen,
(b) outside the maisonette.

9. Compile the nodes-regions incidence matrix for each of the networks in Fig. 26.9. Are the networks topologically equivalent? Do they have the same arcs-nodes incidence matrices? Can you find two non-equivalent networks with the same arcs-regions incidence matrices?

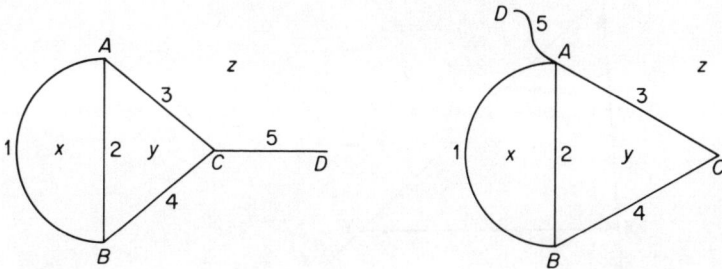

FIG. 26.9

129

# Chapter 27
# Geometry

*Vector geometry*

If $A$ and $B$ are two points, the directed line segment $AB$ can be considered as a vector **AB**. If $\mathbf{AB} = \mathbf{XY}$ the line segments $AB$ and $XY$ are equal in length and parallel. This fact is useful in establishing many geometrical results. For example, in Fig. 27.1 $X$ and $Y$ are the mid-points of the sides $AB$ and $AC$ of the triangle $ABC$. We have

$$\mathbf{AB} + \mathbf{BC} = \mathbf{AC} \Rightarrow \mathbf{BC} = \mathbf{AC} - \mathbf{AB}.$$

Also 
$$\mathbf{XY} = \mathbf{AY} - \mathbf{AX} \Rightarrow \mathbf{XY} = \tfrac{1}{2}\mathbf{AC} - \tfrac{1}{2}\mathbf{AB}$$

since $\mathbf{AY} = \tfrac{1}{2}\mathbf{AC}$ and $\mathbf{AX} = \tfrac{1}{2}\mathbf{AB}$. Hence

$$\mathbf{XY} = \tfrac{1}{2}(\mathbf{AC} - \mathbf{AB}) \Rightarrow \mathbf{XY} = \tfrac{1}{2}\mathbf{BC}.$$

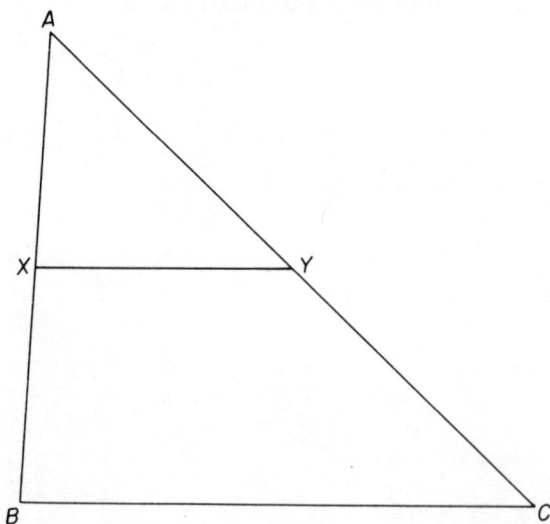

FIG. 27.1

This single vector equation yields the two geometrical results

  (i) $XY = \frac{1}{2}BC$,

  (ii) $XY$ is parallel to $BC$.

*Loci*

The locus of a point subject to some condition is the set of possible positions of the point. Three important plane loci are:

  (a) $\{P:AP = PB\}$, the set of points equidistant from two fixed points $A$ and $B$, is the mediator of $AB$ (see Fig. 27.2).

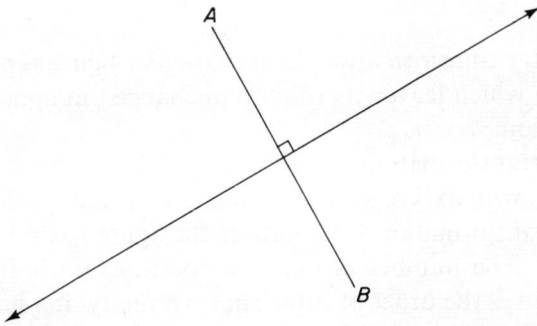

FIG. 27.2

  (b) $\{P:P$ is equidistant from lines $l$ and $l'\}$ consists of the bisectors of the angles formed by $l$ and $l'$ (see Fig. 27.3).

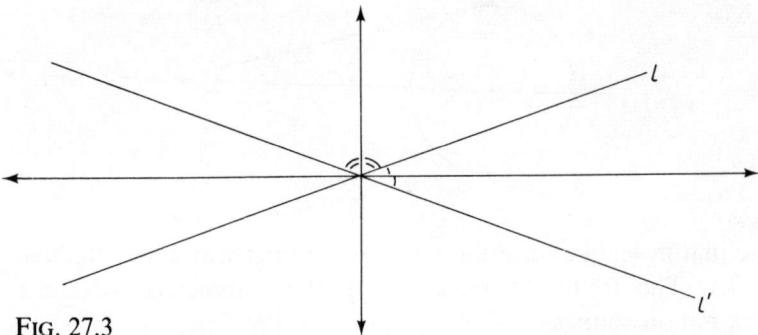

FIG. 27.3

131

(c) $\{P : A\hat{P}B = \text{constant}\}$, the set of points at which the line $AB$ subtends a fixed angle, consists of two arcs of two circles (see Fig. 27.4). If the constant angle is 90° the locus is a complete circle with $AB$ as diameter.

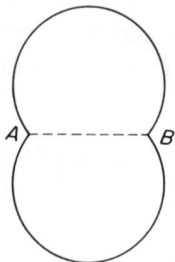

FIG. 27.4

*Symmetry*

A symmetry transformation for a particular figure is a transformation which leaves that figure unchanged in appearance and position.

If the transformation is a reflection the figure has line symmetry, with axis of symmetry the mirror line.

If the transformation is a rotation, the figure has rotational symmetry. The number of different rotations (including the identity) gives the order of rotational symmetry. If a half turn is included in the set of symmetries, the figure has point symmetry about the centre of rotation.

Symmetry properties of figures can give useful geometrical information, e.g. in Fig. 27.5 the tangents $AB$ and $AC$ are

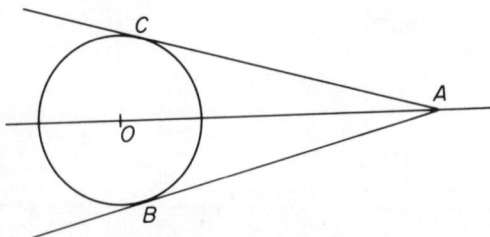

FIG. 27.5

equal in length since the figure is symmetrical about the line $OA$. The following table shows the symmetries of some common shapes.

| Figure | Symmetries | | |
|--------|------|----------|-------|
| | *Line* | *Rotation* | *Point* |
| Circle. | About any diameter. | Through any angle about centre. | About centre. |
| Isosceles triangle | About *AD*. | Identity only. | No. |
| Equilateral triangle | About *AD*, *BE* and *CF*. | About *X* through angles 0°, 120° and 240°. Order 3. | No. |
| Kite | About one diagonal, *AC*. | Identity only. | No. |
| Parallelogram | None. | About *X* through angles 0° and 180°. Order 2. | About *X*. |
| Rhombus | About diagonals *AC* and *BD*. | About *X* through angles 0° and 180°. Order 2. | About *X*. |
| Rectangle | About mediators of sides, *l* and *m*. | About *X* through angles 0° and 180°. Order 2. | About *X*. |

| Figure | Symmetries | | |
| --- | --- | --- | --- |
| | Line | Rotation | Point |
| Square<br> | About diagonals and mediators of sides. | About $X$ through angles $0°$, $90°, 180°$ and $270°$. Order 4. | About $X$. |

## EXERCISES

1. Describe the symmetries of each of the shapes in Fig. 27.7.

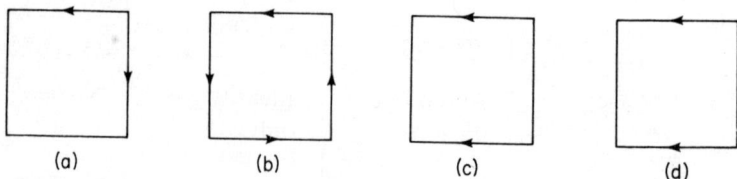

(a)  (b)  (c)  (d)

FIG. 27.7

2. Describe the planes of symmetry of the following objects:

(a) a matchbox,  (b) a cup,
(c) a regular tetrahedron,  (d) a rugby football.

3. The image of the unit square, whose vertices are $O(0,0)$, $A(1,0)$, $B(1,1)$ and $C(0,1)$, under a certain transformation has line symmetry about $y = x$. If the image of $C$ under the transformation has coordinates $(1,3)$, find the coordinates of the image of $A$. Hence write down the matrix of the transformation and find the coordinates of the image of $B$. Describe fully the symmetries of the image.

4. If $A$ and $B$ are fixed points, describe the locus of a point $P$ if the area of triangle $APB$ is fixed.

134

5. $A$ and $B$ are two points 5 cm apart. If $X = \{P : AP \leqslant 4\,\text{cm}\}$, $Y = \{P : AP \leqslant BP\}$, draw an accurate diagram to show $A \cap B$.

6. A quadrilateral $ABCD$ has sides $AB$ and $DC$ equal in length and parallel. If $\mathbf{AB} = \mathbf{x}$ and $\mathbf{BD} = \mathbf{y}$ express $\mathbf{AD}$ and $\mathbf{BC}$ in terms of $\mathbf{x}$ and $\mathbf{y}$ and show that the quadrilateral is a parallelogram.

7. In Fig. 27.8 $O$ is the centre of the circle. Show that $A\hat{O}B = 2A\hat{C}B$. Hence find $A\hat{P}B$ if $A\hat{O}B$ is 124°.

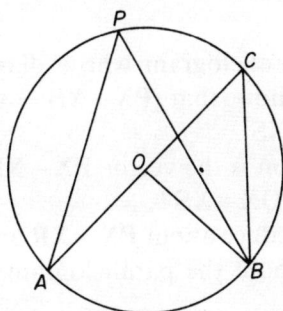

FIG. 27.8

8. Fig. 27.9 shows a circle, centre $O$, with a tangent $ATB$. State the axis of symmetry of the figure and deduce the angle between the tangent and the radius $OT$.

If $P$ is a point on the circle such that $P\hat{T}A = \theta$, deduce the size of $P\hat{O}T$ in terms of $\theta$.

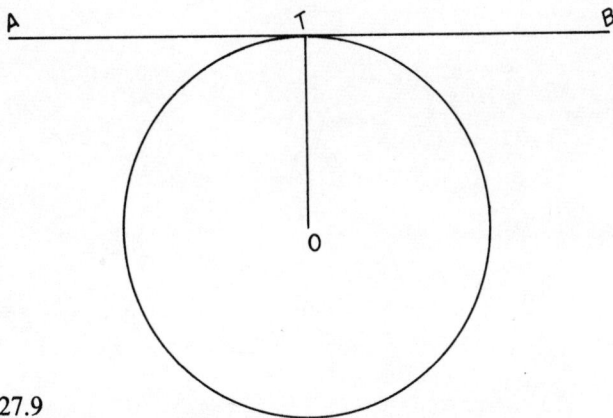

FIG. 27.9

135

9. Describe the following loci in three dimensions:

   (a) $\{P: AP = k\}$, where $A$ is a fixed point and $k$ is constant,
   (b) $\{P: AP = PB\}$, where $A$ and $B$ are fixed points,
   (c) $\{P:$ the distance of $P$ from a fixed line $l$ is constant$\}$.

10. The point $A$ has position vector $\mathbf{a} = \begin{pmatrix} -1 \\ 2 \end{pmatrix}$, and $\mathbf{b} = \begin{pmatrix} 2 \\ 1 \end{pmatrix}$.
Show on a diagram the points whose position vectors are $\mathbf{a} - \mathbf{b}$, $\mathbf{a} + \mathbf{b}$, $\mathbf{a} + 2\mathbf{b}$ and $\mathbf{a} - 2\mathbf{b}$. Hence describe the locus $\{P: \mathbf{OP} = \mathbf{a} + t\mathbf{b}$, $t$ is a real number$\}$.

11. $PQRS$ is a parallelogram whose diagonals $PR$ and $QS$ intersect at $X$. Show that $\mathbf{PX} + \mathbf{XS} = \mathbf{QX} + \mathbf{XR}$ (i.e. that $\mathbf{PX} - \mathbf{XR} = \mathbf{QX} - \mathbf{XS}$).

   In which direction is the vector $\mathbf{PX} - \mathbf{XR}$? Is this the same as the direction of $\mathbf{QX} - \mathbf{XS}$?

   What can you deduce about $\mathbf{PX} - \mathbf{XR}$ and $\mathbf{QX} - \mathbf{XS}$? What geometrical fact about the parallelogram have you proved?

# Chapter 28
# Structure

In arithmetic modulo $n$, a number is represented by its remainder on division by $n$, e.g. in arithmetic modulo 7,

$4+5 = 2$, since  9 leaves remainder 2 on division by 7,
$6 \times 4 = 3$, since 24 leaves remainder 3 on division by 7.

The tables below show the operation of multiplication modulo 6 on $\{1, 2, 3, 4, 5\}$, and addition mod. 5 on $\{0, 1, 2, 3, 4\}$ respectively.

| × | 1 | 2 | 3 | 4 | 5 | | + | 0 | 1 | 2 | 3 | 4 |
|---|---|---|---|---|---|---|---|---|---|---|---|---|
| 1 | 1 | 2 | 3 | 4 | 5 | | 0 | 0 | 1 | 2 | 3 | 4 |
| 2 | 2 | 4 | 0 | 2 | 4 | | 1 | 1 | 2 | 3 | 4 | 0 |
| 3 | 3 | 0 | 3 | 0 | 3 | | 2 | 2 | 3 | 4 | 0 | 1 |
| 4 | 4 | 2 | 0 | 4 | 2 | | 3 | 3 | 4 | 0 | 1 | 2 |
| 5 | 5 | 4 | 3 | 2 | 1 | | 4 | 4 | 0 | 1 | 2 | 3 |

Addition mod. 5 is closed on $\{0, 1, 2, 3, 4\}$ since combining any two of 0,1,2,3,4 gives a member of $\{0, 1, 2, 3, 4\}$. Multiplication mod. 6 is not closed on $\{1, 2, 3, 4, 5\}$ since $2+4 = 0$, and $0 \notin \{1, 2, 3, 4, 5\}$.

1 is the identity element for multiplication mod. 6, since multiplication of any number by 1 has no effect on that number. Similarly 0 is the identity element for addition mod. 5.

In the second table the inverse of 2 is 3, since $3+2 = 2+3 = 1$, and 1 is the identity element. Inspection of the table reveals that every element has an inverse. In the first table 1, 2, 3 and 4 have no inverses, whilst 1 and 5 are self-inverse.

An operation is commutative if the result of combining two elements is unchanged by interchanging the two elements, e.g.

multiplication is commutative since $a \times b = b \times a$,
subtraction is not commutative since in general $a - b \neq b - a$.

Both operations in the tables are commutative. This can be verified by observing that both operation tables are symmetrical about the leading diagonal.

An operation is associative if an expression involving three elements can be written non-ambiguously without brackets, e.g.

addition is associative since $a + (b + c) = (a + b) + c$,
division is not associative since in general $a \div (b \div c) \neq (a \div b) \div c$.

*Boolean Algebra*

When considering switching circuits, the statement $p = 1$ indicates that the switch $p$ is on, $p = 0$ indicates that the switch $p$ is off. The switch $p'$ always takes the opposite value of the switch $p$.

Fig. 28.1

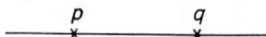

The circuit in Fig. 28.1 is represented by $p \cdot q$, that in Fig. 28.2 by $p + q$. Complicated circuits can be represented in terms of these two operations, and simplified by using the

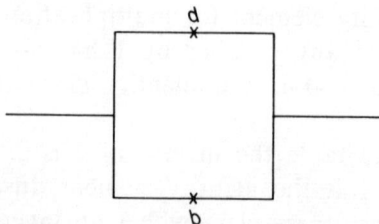

Fig. 28.2

following properties of Boolean Algebra, which can be verified by considering the appropriate circuits.

$$p.p = p+p = p$$

$$p.1 = p \qquad\qquad p+0 = p$$
$$p.0 = 0 \qquad\qquad p+1 = 1$$
$$p+q.r = (p+q).(p+r) \qquad p.(q+r) = p.q+p.r$$
$$p.p' = 0 \qquad\qquad p+p' = 1$$
$$(p.q)' = p'+q' \qquad\qquad (p+q)' = p'.p'$$

The operations . and + are commutative and associative.

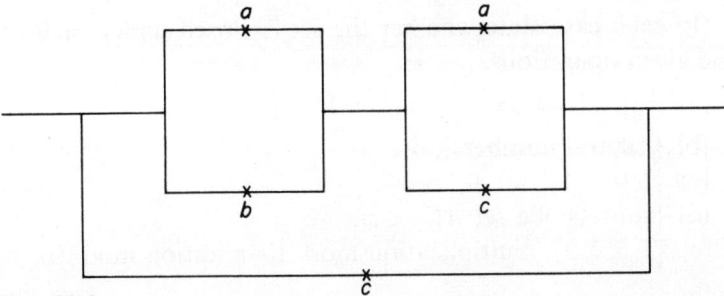

FIG. 28.3

*Example* The circuit in Fig. 28.3 can be represented by the Boolean function

$$(a+b).(a+c)+c = a+b.c+c$$
$$= a+c.(1+b)$$
$$= a+c$$

Thus the circuit is equivalent to that in Fig. 28.4.

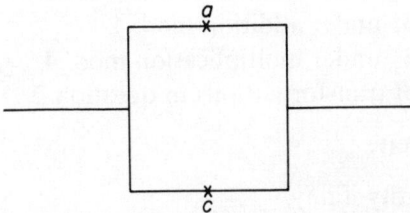

FIG. 28.4

If $p$ and $q$ represent statements the same rules apply if . (in this case sometimes written $\wedge$ ) is taken to mean 'and', and + (sometimes written $\vee$ ) is taken to mean 'or'. Here $p'$ (sometimes written $\sim p$) represents the statement 'not $p$', e.g. if

139

$$p = \text{I like cabbage,}$$
$$q = \text{I like carrots,}$$

then $\quad (p \wedge q) \vee (\sim p) = $ either I like cabbage and carrots or I do not like cabbage.

## EXERCISES

1. In each case state whether the set is closed under each of the given operations.

   (a) {integers}, $\times$, $-$, $\div$, $+$,
   (b) {natural numbers}, $\times$, $\div$, $-$, $+$,
   (c) $\{1, 0, -1\}$, $\times$, $+$, $\div$, $-$,
   (d) {subsets of a set $A$}, $\cap$, $\cup$,
   (e) $\{1, 3, 7, 9\}$, multiplication mod. 10, addition mod. 10.

2. If $\mathbf{X}$ = reflection in $y$-axis,
     $\mathbf{R}$ = rotation through 180° about (0, 0),
which elements must be added to the set $A = \{\mathbf{X}, \mathbf{R}\}$ if it is to be closed under the operation of combining transformations? What polygon has this as its set of symmetries?

3. Construct operation tables for

   (a) $\{1, 2, 3, 4\}$ under multiplication mod. 5,
   (b) $\{0, 1, 2, 3\}$ under addition mod. 4,
   (c) $\{0, 1, 2, 3\}$ under multiplication mod. 4,
   (d) the set of transformations in question 2.

In each case state

   (i) the identity if any,
   (ii) the inverse of each element.

4. If $a$ and $b$ are numbers, investigate whether the operation $*$ is   (a) associative    (b) commutative, if

(i) $a * b = a + b + 2$,   (ii) $a * b = a + 2b$,   (iii) $a * b = 2a + 2b$.

5. Simplify the following Boolean expressions.

(a) $p + p.q + p'.q$,

(b) $p.q.r + p.(q + p')$,

(c) $(p + q).(q + r) + p.r'$,

(d) $(p + q)'.(p.q)$.

6. Write down the Boolean function for each of the circuits in Fig. 28.5. Hence simplify the circuits.

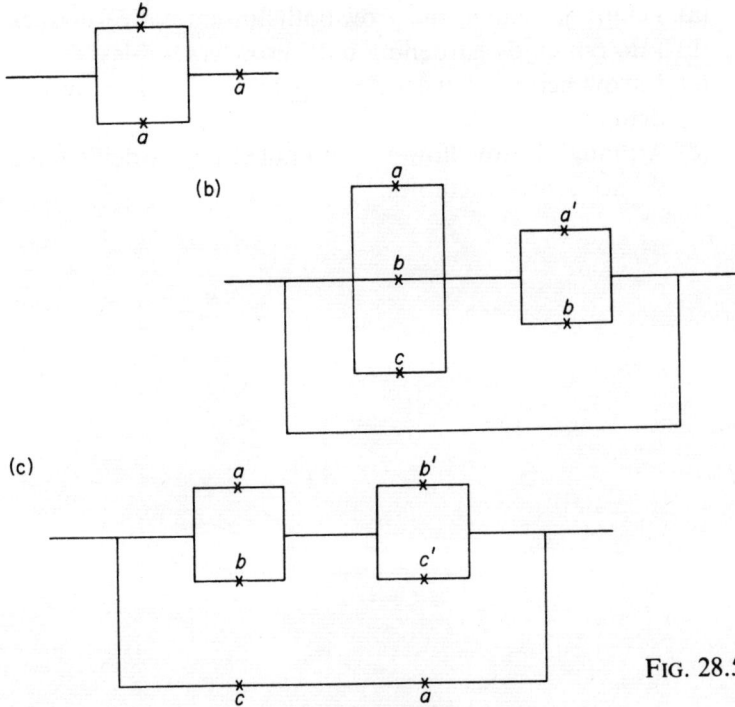

FIG. 28.5

7. If $p$, $q$ and $r$ represent the statements

I use my car,
I travel by bus,
I am sometimes late,

respectively, express the following in English:

(a) $(p \wedge q) \vee r$,

(b) $(\sim q) \wedge p$,

(c) $p \vee (q \wedge r)$,

(d) $(p \wedge \sim r) \vee (q \wedge r)$.

141

8. If *a*, *b* and *c* represent the statements

> I grow flowers,
> I grow vegetables,
> I enjoy gardening,

respectively, write down Boolean expressions corresponding to each of the following statements:

   (a) I enjoy gardening and grow both flowers and vegetables.
   (b) I do not enjoy gardening but I grow vegetables.
   (c) I grow neither flowers nor vegetables, but I enjoy gardening.
   (d) Although I grow flowers, I do not enjoy gardening and do not grow vegetables.

# Miscellaneous Exercises

*Exercise A*

1. Give the coordinates of the image of the point $(8, 3)$ after a quarter turn about the origin followed by reflection in $y = 0$.

2. For which value of $k$ has the matrix

$$A = \begin{pmatrix} 2 & k \\ 3 & 5 \end{pmatrix}$$

no inverse?

If $k = 2$, find the inverse of $A$.

Find the simultaneous solution of the equations

$$4x + y = 2,$$
$$3x + 2y = 1.$$

4. Find the prime factors of $10^2 - 7^2$.

5. Express $(4\frac{1}{3} - 2\frac{1}{4})/1\frac{2}{3}$ in its simplest form.

6. In nine innings a batsman scores

$$16, 38, 0, 5, 24, 73, 12, 30, 52.$$

If he was out at the end of each innings calculate:

(a) his mean score,    (b) his median score.

7. Estimate to one significant figure the value of $\sqrt{(810 \times 27.3)/0.0042}$, giving your answer in standard index form.

8. If **X** is a direct transformation and **Y** is an indirect transformation, for which values of $m$ is $\mathbf{X}^2\mathbf{Y}^m$ a direct transformation?

9. Find the sum of the angles of a septagon and deduce the angle between two sides of a regular septagon.

10. Draw a Venn diagram to represent the sets $A$, $B$ and $C$ if

(a) $A \subset B$ and $A \cap C \neq \emptyset$,     (b) $A \cap B \subset C$.

*Exercise B*

1. Find the prime factors of $273_9$, giving your answers in base nine.

2. The point $A'$ is the image of the point $A$ under the translation $\begin{pmatrix} -2 \\ 3 \end{pmatrix}$ and the point $A^*$ is the image of $A$ under the translation $\begin{pmatrix} 1 \\ -2 \end{pmatrix}$. Give the column vector of the translation which maps $A^*$ onto $A'$.

3. Find the gradient of the line which passes through $(-1, 0)$ and $(1, 3)$. Does this line pass through $(-3, 3)$?

4. Solve the equation $\frac{1}{3}(3-2x) = 3$. Hence simplify $\{x: \frac{1}{3}(3-2x) < 3\}$.

5. Find the values of $x$ and $y$ if

$$x + y = 5$$
$$x^2 - y^2 = 12.$$

6. The probability that I remember to do my maths homework is $\frac{2}{3}$, and that I remember to do my English homework is $\frac{3}{4}$. What is the probability that on a night when the two subjects are set I forget to do both?

7. If $x * y = \sqrt{(x+y)}$, calculate $(4 * 5) * 7$ and $4 * (5 * 7)$. What can you deduce about the operation $*$?

8. Compile the direct route matrix for the network in Fig. E.1.

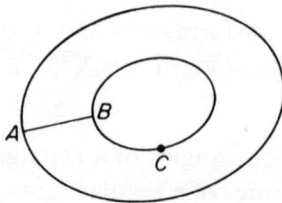

FIG. E.1

By calculating the two-stage matrix for this network, state which two points are not connected by a two-stage route.

9. A rhombus has diagonals of lengths 24 cm and 16 cm. Find the area of the rhombus.

10. Make $t$ the subject of the formula $s = \frac{1}{2}gt^2$. Calculate the value of $t$ when $s = 14\cdot4$ and $g = 9\cdot8$.

*Exercise C*

1. Represent graphically
$$\{(x, y): x + 2y \leqslant 8\} \cap \{(x, y): y \geqslant 3x - 2\}.$$

2. Calculate the difference in longitude between two points on the earth's surface 1000 nautical miles apart

(a) if both points are on the equator,
(b) if both points have latitude 40° N.

3. A shear has invariant line $y = 0$ and maps $(1, 3)$ onto $(7, 3)$. State the matrix of this shear.

4. A cyclist passes a road junction at 1.00 p.m. travelling at a speed of 15 km/h. One hour later a motorist passes the same junction travelling in the same direction at 60 km/h. Assuming both continue along the same road at a steady speed, at what time will the car overtake the cycle?

5. If $p \propto 1/q$ and $p = 120$ when $q = \frac{1}{2}$, find the value of $p$ when $q = 2$.

6. Two spherical balls are made of the same material and have radii 6 cm and 8 cm. If the larger ball weighs 1 kg, find the weight of the smaller ball correct to two significant figures.

7. An isosceles triangle has two sides of length 24 cm and one angle of 110°. Find the length of the third side.

8. In a class of 30 children, 23 obtain a pass mark in the English exam, 18 a pass mark in the French exam and 15

145

pass both. What is the probability that a child chosen at random from the class failed both subjects?

9. An estimate is made of the value of $\pi$ by taking the area of the circle in Fig. E.2 to be the mean of the areas of the two squares. What value does this give for $\pi$?

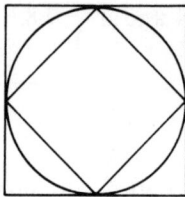

FIG. E.2

10. The relation $f$ is defined on the set of $2 \times 2$ matrices by

$$\mathbf{A} \to \text{the determinant of } \mathbf{A}.$$

(a) Is $f$ a function?
(b) Is the inverse of $f$ a function?
(c) What can you say about $\mathbf{A}$ if $f(\mathbf{A}) = 0$?

*Exercise D*
In each question state whether the given statement is true or false. If the statement is false, modify it to make it true.

1. $A \cap C' = \varnothing \Leftrightarrow A \subset C$.

2. $\{x : x^2 - 9 = 0\} \cap \{x : x < 0\} = \varnothing$.

3. If A, B and C are non-singular matrices of order $2 \times 2$,

$$\mathbf{AB} = \mathbf{C} \Rightarrow \mathbf{B} = \mathbf{A}^{-1}\mathbf{C}.$$

4. If $f$ and $g$ are functions the inverse of $fg$ is $f^{-1}g^{-1}$.

5. $\mathbf{AB} = \mathbf{XY} \Rightarrow Y$ is the image of $B$ under the translation $\mathbf{AX}$.

6. Rotation through $90°$ about $(0,0)$ followed by reflection in $x = 0$ is equivalent to reflection in $y = x$.

7. The transformation with matrix $\begin{pmatrix} 2 & 0 \\ 0 & 2 \end{pmatrix}$ changes the area of a figure by a factor of 2.

8. The equations $2y = 4x + 3$ and $y - 2x = 1\frac{1}{2}$ have no simultaneous solution.

9. The value of 7149 to two significant figures is 7200.

10. $3x - (2x + 1)(3 - x) = 2x^2 - 2x - 3$.

11. $y \propto 1/x \Leftrightarrow x \propto 1/y$.

12. If the probability of a child being a boy is $\frac{1}{2}$, the probability of just two children in a family of four children being boys is $\frac{1}{2}$.

13. The area between two concentric circles of radii $a$ and $b$ $(a > b)$ is $\pi(a - b)(a + b)$.

14. The point on the earth's surface diametrically opposite the point (latitude $40°$ S, longitude $130°$ W) is the point (latitude $40°$ N, longitude $130°$ E).

15. In the set of real numbers the multiplicative inverse of 0 is 0.

In exercises E and F the correct answer should be chosen from the alternatives given.

*Exercise E*
1. If $a * b = \frac{1}{2}(a + b)$, the operation $*$ is

   (a) associative and commutative,
   (b) commutative but not associative,
   (c) associative but not commutative,
   (d) neither associative nor commutative.

2. A shear has invariant line $y = x$ and maps $(-3, -1)$ onto $(-2, 0)$. The image of $(5, 1)$ under this shear is

      (a) $(7, 3)$,   (b) $(6, 2)$,   (c) $(4, 0)$,   (d) $(3, -1)$.

3. After five mathematics tests, each marked out of 10, my average mark is 7. On the sixth test I score 9 out of 10. My mean mark becomes

      (a) $7\frac{1}{3}$,   (b) $7\frac{1}{2}$,   (c) 8,   (d) none of these.

147

4. The inverse of the matrix $\begin{pmatrix} -2 & 1 \\ -5 & 3 \end{pmatrix}$ is

(a) $\begin{pmatrix} 3 & -1 \\ 5 & -2 \end{pmatrix}$,

(b) $\begin{pmatrix} -3 & 1 \\ -5 & 2 \end{pmatrix}$,

(c) $\begin{pmatrix} -3/11 & 1/11 \\ -5/11 & 2/11 \end{pmatrix}$,

(d) $\begin{pmatrix} 3 & 5 \\ -1 & -2 \end{pmatrix}$.

5. The equation $(x+2)^2 = x(x+3)-2$ has

(a) no solutions,     (b) one solution,
(c) two solutions,     (d) infinitely many solutions.

6. If **S** denotes reflection in $x = 2$ and **T** denotes reflection in $x = 6$ then **ST** is a

(a) translation,     (b) rotation,
(c) reflection,     (d) glide-reflection.

7. A polygon has exactly two axes of symmetry. The polygon can be a

(a) parallelogram,   (b) square,   (c) triangle,   (d) rhombus.

8. If $\sin \theta° = \frac{3}{5}$ $(90 < \theta < 270)$, then $\tan \theta° =$

(a) $-1\frac{1}{3}$,   (b) $-\frac{3}{4}$,   (c) $1\frac{1}{3}$,   (d) $\frac{3}{4}$.

9. If $n(A) = 15$, $n(B) = 23$ and $n(A \cap B) = 7$ then $n(A \cup B) =$

(a) 38,   (b) 31,   (c) 45,   (d) 24.

10. The value of $\sqrt{(276/47)}$ to one significant figure is

(a) 0·1,   (b) 0·3,   (c) 3,   (d) 0·4.

*Exercise F*
1. If $A$ is a matrix of order $3 \times 2$ and $B$ a matrix of order $3 \times 5$ which of the following matrix products is compatible?

(a) $AB$,   (b) $BA$,   (c) $AB'$,   (d) $B'A$.

2. I miss the bus on my way to work with probability $\frac{1}{8}$, and on the way home with probability $\frac{1}{3}$. The probability that I miss the bus to work but catch it on the way home is

(a) $\frac{19}{24}$, (b) $\frac{1}{12}$, (c) $\frac{5}{24}$, (d) $\frac{1}{6}$.

3. A transformation maps $(1,0)$ onto $(3,2)$ and $(1,1)$ onto $(2,5)$. The matrix of the transformation is

(a) $\begin{pmatrix} 3 & 2 \\ 2 & 5 \end{pmatrix}$, (b) $\begin{pmatrix} 3 & 1 \\ 2 & 5 \end{pmatrix}$,

(c) $\begin{pmatrix} 3 & -1 \\ 2 & 3 \end{pmatrix}$, (d) $\begin{pmatrix} 3 & 2 \\ -1 & 3 \end{pmatrix}$.

4. The gradient of the line $2y + 3x = 7$ is

(a) $-1\frac{1}{2}$, (b) $1\frac{1}{2}$, (c) $-3$, (d) $\frac{2}{3}$.

5. In a triangle $ABC$, $\mathbf{AB} = \mathbf{b}$ and $\mathbf{AC} = \mathbf{c}$, $\mathbf{BC}$ is equal to

(a) $\mathbf{b} + \mathbf{c}$, (b) $\mathbf{b} - \mathbf{c}$, (c) $\mathbf{c} - \mathbf{b}$, (d) $\frac{1}{2}(\mathbf{b} + \mathbf{c})$.

6. The inverse of the function $x \to (4/x) - 5$ is

(a) $x \to \dfrac{4}{x} + 5$, (b) $x \to 4x + 5$,

(c) $x \to \dfrac{4}{x+5}$, (d) $x \to \dfrac{4}{x} + \dfrac{1}{5}$.

7. If

$$\frac{3 \times 10^4}{4 \times 10^{-2}} = 7 \cdot 5 \times 10^n,$$

then the value of $n$ is

(a) 7, (b) 5, (c) 6, (d) 1.

8. The table

| $x$ | 2 | 3 | 4 |
|---|---|---|---|
| $y$ | 15 | 10 | 7·5 |

149

is compatible with the statement

(a) $y \propto x$, (b) $y \propto 1/x$, (c) $y \propto x^2$, (d) $y \propto 1/x^2$.

9. I have 30p and buy $y$ apples at 2p each and $x$ pears at 3p each. This information can be written

(a) $3x + 2y < 30$, (b) $x \leqslant 10$, $y \leqslant 15$,
(c) $3x + 2y = 30$, (d) $3x + 2y \leqslant 30$.

10. The radius of a circle equal in area to a square of side $a$ is

(a) $a^2/2\pi$, (b) $a/\sqrt{\pi}$, (c) $a\sqrt{\pi}$, (d) $a/\pi$.

*Exercise G*

1. The transformation with matrix $\begin{pmatrix} 0 & -1 \\ 1 & 2 \end{pmatrix}$ is applied to the quadrilateral $OABC$, where $O, A, B$ and $C$ are the points $(0,0)$, $(-3,3)$, $(-2,4)$ and $(1,1)$ respectively. Show on the same diagram the quadrilateral $OABC$ and its image under this transformation. Describe the transformation geometrically.

2. A bag contains 5 blue discs and 3 red discs, which are drawn at random and not replaced. Find the probability that

(a) the first disc drawn is red,
(b) the first two discs drawn are different colours,
(c) the third disc drawn is red,
(d) the last disc drawn is blue.

3. A ball is thrown vertically upwards. Its height, $s$ metres, after a time $t$ seconds in the air, is given by

$$s = 40t - 5t^2.$$

Draw a graph to represent the height of the ball for all relevant values of $t$ and use this graph to find

(a) the total time of flight,
(b) the maximum height attained,
(c) the speed of the ball after it has been in the air for 2 seconds.

4. Use Venn diagrams to prove that

$$(A \cap B) \cup C' = [(A' \cup B') \cap C]'$$

5. Construct the nodes-arcs incidence matrix for a cube and draw the plane network corresponding to this incidence matrix.

6. Express as fractions in their simplest terms

(a) $0 \cdot \dot{7}$,     (b) $0 \cdot \dot{2}\dot{7}$,     (c) $0 \cdot 8\dot{4}\dot{0}$.

7. A manufacturer raises the price of an article by 5%. If at the old price he made a profit of 10% of the price, and 75% of the price increase is taken by increased costs of manufacture, find the new profit as a percentage of the new price.

8. Two cylinders, each of length 1·6 m and radius 3 cm are laid side by side with a third similar cylinder placed on top of them, (see Fig. E.3). Calculate the volume of the space between the cylinders.

9. Find the angle between the hands of a clock at 6.15.

FIG. E.3

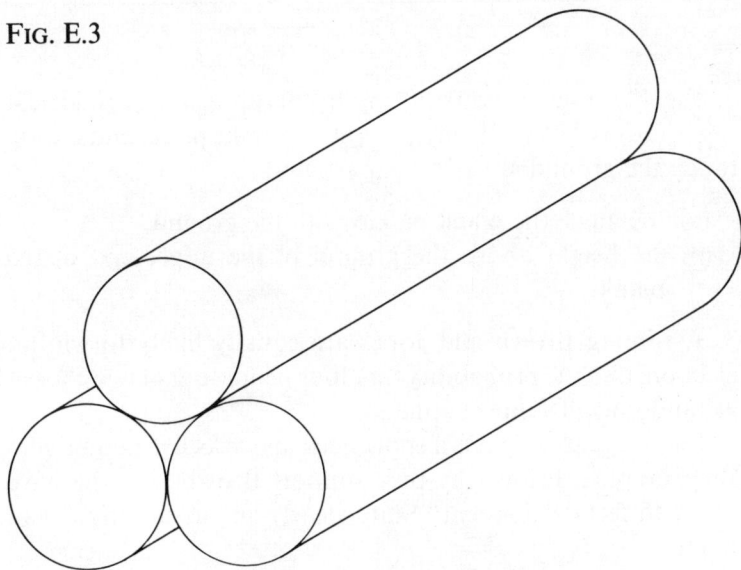

10. Find the coordinates of the vertices of the triangle formed by the lines $3x + 2y = 2$, $4x = 3y + 7$ and $x = 3$.

*Exercise H*
1. If the points $A$ and $B$ have coordinates $(1, 5)$ and $(7, 2)$ respectively, describe **AB** as a column vector. If $M$ is the mid-point of $AB$ find **OM** in terms of **OA** and **AB**, and hence state the coordinates of $M$. Similarly find the point on the line $AB$ such that $AN/NB = \frac{1}{2}$.

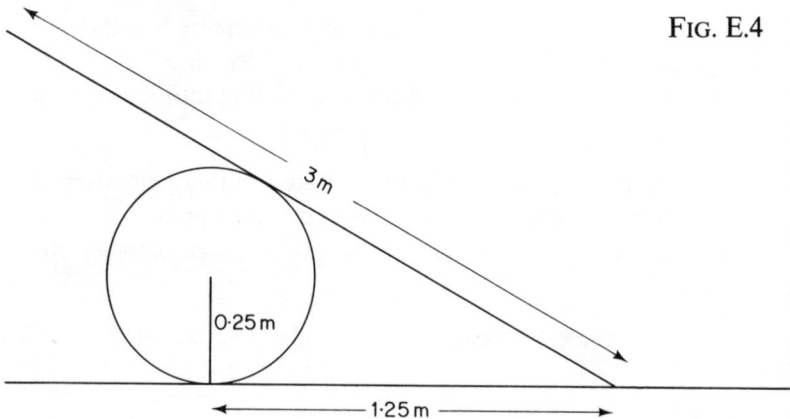

FIG. E.4

2. Fig. E.4 shows a plank of length 3 m resting on a cylindrical log of radius 0·25 m. If the points at which the plank and the log touch the ground are 1·25 m apart, find

(a) the angle the plank makes with the ground,
(b) the height above the ground of the upper end of the plank.

3. Assuming Brown and Jones are equally likely to win an election, find the probability that four people out of five chosen at random will support Jones.

Jones' agent conducts a spot check and asks five people who they support. If four say they support Brown, does he have cause to feel despondent? State clearly any assumptions you make.

4. The front page of a newspaper is in the shape of a rectangle 40 cm by 60 cm and contains news, advertisements and photographs. The editorial policy is to have at least twice as much space devoted to news as advertisements, but at least 250 cm$^2$ of advertising space is essential to maintain the advertising revenue. Use a graphical method to determine the greatest space that can be allocated to

    (a) photographs,   (b) news,   (c) advertisements.

5. It is suspected that variables $q$ and $s$ are connected by a relation of the form $q = (a/s) + b$, where $a$ and $b$ are constants. An experiment yields the following results:

| $s$ | 2·0 | 3·0 | 4·0 | 5·0 | 6·0 |
|---|---|---|---|---|---|
| $q$ | 1·4 | 0·83 | 0·54 | 0·33 | 0·25 |

By drawing a suitable graph say whether you consider the results to be consistent with the relation $q = (a/s) + b$, and if so state suitable values of $a$ and $b$.

6. Solve the matrix equation

$$\begin{pmatrix} 1 & 3 \\ 0 & 1 \end{pmatrix} \mathbf{X} = \begin{pmatrix} 1 & -1 \\ 0 & 1 \end{pmatrix}.$$

Interpret your answer in terms of the transformations represented by the matrices.

7. A man collecting eggs from his chickens classifies them as large or small and as brown or white. One day he finds that five large eggs, four brown eggs and seven white eggs are cracked. How many cracked eggs does he collect? In all he collects 56 large brown eggs, 72 small brown eggs and 105 large eggs. If he collects 183 eggs altogether, how many large brown eggs are cracked?

8. If $A$ = {regular polygons}, $B$ = {polygons containing at least one right angle} and $\xi$ = {triangles}, describe as concisely as possible $A$, $B$ and $A \cap B$.

153

For which value or values of $n$ is $A \cap B = \emptyset$ if $\xi = \{n\text{-sided}$ polygons$\}$?

9. State the gradient of the line whose equation is $y = mx + c$. Make $m$ the subject of the formula $y = mx + c$ and give a geometrical interpretation of your result by considering a point $(x, y)$ on the line and the point where the line cuts the $y$-axis.

10. State the number of elements in the solution sets of

(a) $(x+2)(x-3) = x^2 + 5x - 1$,  (b) $(x-3)^2 - 5 = x^2 - 6x + 1$,
(c) $(x+3)^2 = x^2 + 9$,  (d) $1/(x-2) = 0$.

*Exercise I*
1. 1000 screws are taken as a sample of the output of a machine, and their lengths measured to give the following results.

| Length (mm) | 27·0– 27·5 | 27·5– 28·0 | 28·0– 28·5 | 28·5– 29·0 | 29·0– 29·5 | 29·5– 30·0 | 30·0– 30·5 |
|---|---|---|---|---|---|---|---|
| No. of screws of this length | 3 | 52 | 239 | 411 | 228 | 59 | 8 |

Draw a cumulative frequency curve to represent this information. If screws are only accepted if they lie within the range 28·25 mm–29·75 mm, use your graph to estimate the percentage of screws rejected.

2. Construct the image of the unit under the transformation with matrix

$$A = \begin{pmatrix} 1 & -1 \\ 1 & 1 \end{pmatrix}$$

and describe the geometrical effect of the transformation. Similarly, describe the effect of the transformation with matrix

$$B = \begin{pmatrix} 3 & 1 \\ 2 & 1 \end{pmatrix}.$$

By considering solutions of the equation

$$\mathbf{BA}\begin{pmatrix} x \\ y \end{pmatrix} = \begin{pmatrix} x \\ y \end{pmatrix},$$

investigate the existence of an invariant line for the transformation **BA**.

3. If $f$ and $g$ are the functions $x \to 3x+2$ and $x \to 4x-1$ respectively state the gradients of the graphs of $f$, $g$ and $fg$. What is the connection between these three quantities?

By considering the functions $x \to ax+b$ and $x \to cx+d$ generalize your result.

4. In an election for school captain, each of the 50 members of the senior year has one vote and there are three candidates: Alan, Brian and Charles. In a survey of voting intentions each voter was asked to write down the candidate or candidates he was considering voting for. Twelve pupils mentioned Brian only or mentioned both Brian and Charles but not Alan. Fifteen mentioned Alan only or mentioned both Alan and Charles but not Brian. The numbers of pupils who mentioned Alan and Brian, Brian and Charles or Alan and Charles (in no case all three) were equal. The number who mentioned Charles only was twice the number of those who mentioned all three candidates. If 18 pupils mentioned Charles and 22 mentioned Alan find

(a) the number of pupils undecided between all three candidates,
(b) the number of pupils who mentioned Brian,
(c) the number of pupils intending to abstain if the three candidates are not allowed to vote.

5. List the elements of the set

$$A = \{x : (3x+4)(x-2)(x^2-5) = 0\}$$

when (a) $\xi = \{\text{integers}\}$, (b) $\xi = \{\text{rational numbers}\}$, (c) $\xi = \{\text{real numbers}\}$.

6. A boat sets sail due North at 20 km/h. The boat is taken off course by a current of 8 km/h flowing from the south-east. Find by scale drawing or otherwise the position of the boat one hour after setting sail.

7. Fig. E.5 shows the cross-section of a tent in the form of an equilateral triangle $PQR$, a vertical pole $BC$ of length 1 m and a source of light at $A$. If $AB = 2$ m and $BP = 1\cdot5$ m, find the length of the shadow of the pole on the tent.

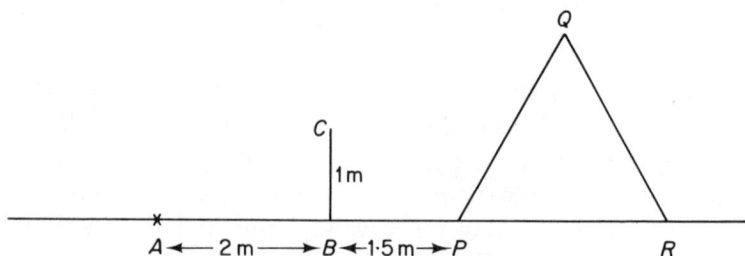

FIG. E.5

8. A coffee shop produces gift packs of blends of coffee. Pack $A$ contains 3 packets of blend $X$, 2 of blend $Y$ and 4 of blend $Z$, and costs 140p. Pack $B$ contains 2 packets of blend $X$, 1 of blend $Y$ and 3 of blend $Z$, and costs 95p. Pack $C$ contains one packet each of blends $X$ and $Y$, and 2 packets of blend $Z$, and costs 65p. Find the cost of one packet of each blend.

9. If $a$ and $b$ are real numbers, use the fact that $(a-b)^2 \geqslant 0$ to prove that the mean of the squares of any two numbers cannot be less than the product of the numbers.

10. If $\mathbf{J} = \begin{pmatrix} 0 & -1 \\ 1 & 0 \end{pmatrix}$, show that $\mathbf{J}^4 = \mathbf{I}$ and interpret this result geometrically.

If $p$ and $q$ are real numbers, simplify $p\mathbf{I} + q\mathbf{J}$ and show that the set of matrices of this form is closed under addition and multiplication.

156

# Answers

## Chapter 1

1. (a) $\{a, f, h\}$,
   (b) $\{a, c, d, f, h, i, j, k, l\}$,
   (c) $\{b, d, e, g, j, m, n\}$,
   (d) $\{b, c, e, g, i, l, m, n\}$,
   (e) $\{b, c, d, e, g, i, j, k, l, m, n\}$,
   (f) $\{d, j\}$,
   (g) $\{a, b, c, e, f, g, h, i, l, m, n\}$,
   (h) $\{a, b, c, d, e, f, g, h, i, j, k, l, m, n\}$,
   (i) $\{a, d, f, h, j, k\}$.

2. (a) $\{a, c, f, h, j, k, m, n\}$,
   (b) $\{a, c, d, f, h, i, j, k, l, m, n\}$,
   (c) $\{a, c, h, m, n\}$,
   (d) $\{c, j, k, m, n\}$,
   (e) $\{d, f, i, l\}$,
   (f) $\{a, c, h, j, k\}$,
   (g) $\{\ \}$.

4. (a) $B$,  (b) $A$,  (c) $\varnothing$.

5. (i) (a) $A$,  (b) $\varnothing$,  (c) $\xi$,  (d) $\varnothing$,  (e) $\varnothing$,  (f) $\varnothing$.
   (ii) (a) true,  (b) true,  (c) false, $A' \cap B' = \varnothing \Rightarrow A \cup B = \xi$,  (d) true.

6. (a) $A \subset B$,  (b) $C \subset B'$,  (c) $B \cap C \subset A$,  (d) $B \cup (A \cap C) = \xi$;
   no houses built in the last five years have two storeys.

7. (a) All members of the chess club are good at maths,
   (b) some members of white house are good at maths,
   (c) all pupils in the school are good at maths or in the chess club,
   (d) all good mathematicians in the chess club are in white house,
   (e) all members of the chess club are in white house,
   (f) all members of the chess club are in white house or are not good mathematicians.

8. $16, 7$.    9. $2$.    10. $1, 7$.    11. $2$.

## Chapter 2

1. (a) $5$,
   (b) $^-4$,
   (c) $^-7$,
   (d) $^-1$,
   (e) $1$,
   (f) $^-12$,
   (g) $^-4$,
   (h) $^-35$,
   (i) $^-6$,
   (j) $^-2\frac{2}{5}$,
   (k) $0$,
   (l) $10$.

2. (a) $^-5$,
   (b) $^-2$,
   (c) $^-1\frac{1}{2}$,
   (d) $8$,
   (e) $^-3$,
   (f) $^-3$.

3. (i) (a) $\{1, 3, 5, 15\}$,        (b) $\{1, 2, 3, 4, 6, 9, 12, 18, 36\}$,

    (c) $\{1, 23\}$,        (d) $\{1, 2, 3, 4, 6, 9, 12, 18, 27, 36, 54, 108\}$,

    (e) $\{1, 2, 4, 8, 16, 32, 64, 128, 256, 512\}$.

  (ii) (a) $\{3, 5\}$,    (b) $\{2, 3\}$,    (c) $\{23\}$,    (d) $\{2, 3\}$,    (e) $\{2\}$.

4. $42 = 2 \times 3 \times 7$, $30 = 2 \times 3 \times 5$, $6$, $210$.

  (a) $6, 420$,    (b) $12, 1320$,    (c) $1, 1537$.

5. $^-5 \in B, C, E$;   $2 \cdot 61 \in C, E$;   $\sqrt{90} \in D, E$;   $0 \cdot 7\dot{1}\dot{6} \in C, E$;   $\pi^2 \in D, E$;   $\sqrt{3} \in D, E$

6. (i) (a) $\varnothing$,    (b) $\{^-2\}$,    (c) $\{^-2\}$.

  (ii) (a) $\varnothing$,    (b) $\{7/3\}$,    (c) $\{7/3\}$.

  (iii) (a) $\{3\}$,    (b) $\{3, ^-3\}$,    (c) $\{3, ^-3\}$.

  (iv) (a) $\varnothing$,    (b) $\varnothing$,    (c) $\{\sqrt{10}, ^-\sqrt{10}\}$.

  (v) (a) $\varnothing$,    (b) $\{^-2\}$,    (c) $\{^-2\}$.

  (vi) (a) $\varnothing$,    (b) $\varnothing$,    (c) $\varnothing$.

7. $1, 3, 5, 7, 9, 11, 13, 15, 17, 19$;      $100$;      $400$;      $225$.

## Chapter 3

1. (a) $\Rightarrow$,   (b) $\Leftarrow$,   (c) not possible,   (d) $\Leftrightarrow$,   (e) $\Leftrightarrow$,

  (f) $\Leftrightarrow$,   (g) $\Leftarrow$,   (h) not possible,   (i) $\Rightarrow$,   (j) not possible.

2. (a) $1024$,   (b) $125$,   (c) $\frac{1}{27}$,   (d) $^-1$,   (e) $1$.

3. (a) $3^7$,   (b) $2$,   (c) $5^{-9}$,   (d) $7^4$,   (e) $8^6$,   (f) not possible,

  (g) $2^5$,   (h) $10^{-2}$,   (i) $5^2$.

4. (a) $2$,   (b) $8$,   (c) $\frac{1}{3}$.

## Chapter 4

1. $\begin{pmatrix} 2 & 5 & 12 & 4 \\ 1 & 8 & 9 & 2 \\ 0 & 13 & 3 & 5 \end{pmatrix} \begin{pmatrix} 10 \\ 4 \\ 1 \\ -1 \end{pmatrix} = \begin{pmatrix} 48 \\ 49 \\ 50 \end{pmatrix}$,

  $A$ 48 points,     $B$ 49 points,     $C$ 50 points.

2. $\begin{pmatrix} 6 & 3 & 3 \\ 0 & 4 & 0 \\ 3 & 0 & 4 \\ 2 & 0 & 0 \\ 6 & 4 & 2 \end{pmatrix} \begin{pmatrix} 2 \\ 3 \\ 2\frac{1}{2} \end{pmatrix} = \begin{pmatrix} 28\frac{1}{2} \\ 12 \\ 16 \\ 4 \\ 29 \end{pmatrix}$,    $89\frac{1}{2}$p.

3. $\begin{pmatrix} 11 & 6 \\ 3 & 2 \\ 13 & 8 \end{pmatrix}$.

4. (a) $\begin{pmatrix} 6 & -3 \\ -9 & 12 \\ -15 & 0 \end{pmatrix}$,  (b) impossible,  (c) $\begin{pmatrix} 2 & -2 \\ 4 & -1 \end{pmatrix}$,

(d) $\begin{pmatrix} -1 & 6 \\ 7 & -14 \end{pmatrix}$,  (e) impossible,  (f) $\begin{pmatrix} -5 & 6 \\ 15 & -14 \\ 5 & -10 \end{pmatrix}$,

(g) $\begin{pmatrix} 7 & -8 \\ 9 & -1 \end{pmatrix}$,  (h) $\begin{pmatrix} 0 & -3 \\ 10 & 2 \\ -10 & 10 \end{pmatrix}$,  (i) $\begin{pmatrix} 2 & -3 & -5 \\ -1 & 4 & 0 \end{pmatrix}$,

(j) $\begin{pmatrix} \frac{1}{2} & \frac{1}{2} \\ \frac{3}{4} & \frac{3}{4} \end{pmatrix}$,  (k) $\begin{pmatrix} \frac{1}{7} & \frac{4}{7} \\ -\frac{1}{7} & \frac{3}{7} \end{pmatrix}$,  (l) $\begin{pmatrix} 5 & -10 & -10 \\ -10 & 25 & 15 \\ -10 & 15 & 25 \end{pmatrix}$.

5. (a) $\begin{pmatrix} 2 & 3 \\ 5 & 8 \end{pmatrix}$,  (b) $\begin{pmatrix} \frac{1}{8} & -\frac{3}{8} \\ \frac{1}{4} & \frac{1}{4} \end{pmatrix}$,  (c) $\begin{pmatrix} \frac{3}{11} & \frac{1}{11} \\ \frac{2}{11} & -\frac{3}{11} \end{pmatrix}$,

(d) singular,  (e) $\begin{pmatrix} -\frac{1}{6} & \frac{1}{2} \\ \frac{1}{3} & 0 \end{pmatrix}$,  (f) $\begin{pmatrix} \frac{3}{a} & -b \\ -\frac{2}{b} & a \end{pmatrix}$.

6. (a) $\mathbf{P+Q} = \mathbf{Q+P} = \begin{pmatrix} 1 & 4 \\ 5 & -5 \end{pmatrix}$, addition is commutative,

(b) $\mathbf{PQ} = \begin{pmatrix} -4 & 13 \\ -7 & 21 \end{pmatrix}$, $\mathbf{QP} = \begin{pmatrix} 13 & -9 \\ -5 & 4 \end{pmatrix}$, multiplication is not

commutative,

(c) $\mathbf{P(Q+R)} = \mathbf{PQ+PR} = \begin{pmatrix} -2 & 18 \\ -3 & 28 \end{pmatrix}$, multiplication is distributive

over addition,

(d) $\mathbf{(PQ)R} = \mathbf{P(QR)} = \begin{pmatrix} -26 & 1 \\ -42 & 0 \end{pmatrix}$, multiplication is associative.

7. $\mathbf{AA'}$ is square.  8. HELP.

9. (a) $\begin{pmatrix} 0 & -6 \\ 0 & -3 \end{pmatrix}$,  (b) $\begin{pmatrix} 9 & 7 \\ -5 & -4 \end{pmatrix}$,  (c) $\begin{pmatrix} -1 & 5 \\ -1 & 6 \end{pmatrix}$,

(d) $\begin{pmatrix} -8 & -10 \\ 5 & 5 \end{pmatrix}$,  (e) $\begin{pmatrix} -2 & -5 \\ -4 & -2 \end{pmatrix}$,  (f) not possible.

159

10. $\begin{pmatrix} ad-bc & 0 \\ 0 & ad-bc \end{pmatrix} = \begin{pmatrix} 0 & 0 \\ 0 & 0 \end{pmatrix},$

(a) $\begin{pmatrix} 4k & -2k \\ -2k & k \end{pmatrix},$  (b) $\begin{pmatrix} 3k & -6k \\ k & -2k \end{pmatrix},$  (c) impossible;

when **A** is singular, infinitely many solutions, no.

## Chapter 5

1. (a) function,    (b) function and inverse function,    (c) function,
   (d) neither,    (e) function,    (f) inverse function.

2. (a) 3,    (b) $-3$,    (c) $-1$,    (d) 11,    (e) $-3$,
   (f) $-4$,    (g) $-7$,    (h) $-28$,    (i) $-9$,    (j) $-22$.

3. (a) $\{-1, -\frac{1}{2}, 0, \frac{1}{2}, 1\}$,    (b) $\{0, 1, 4\}$,    (c) $\{0\}$,
   (d) $\{x : 0 < x \leqslant 1\}$,    (e) $\{x : -1 \leqslant x \leqslant 1\}$,    (f) $\{x : x \geqslant 6\}$.

4. (a) {real numbers},    (b) $\{x : -3 \leqslant x \leqslant 3\}$,
   (c) {real numbers},    (d) $\{x : x \leqslant -4 \text{ or } x \geqslant 4\}$,
   (e) {real numbers except $x = -1$},    (f) $\{x : x > 0\}$,
   (g) {natural numbers > 1}.

5. (a) $x \to \frac{1}{3}(x-4)$,    (b) $x \to 4x+3$,    (c) $x \to 2x+3$,    (d) $x \to \frac{1}{2}(3x+1)$,
   (e) $x \to \frac{1}{4}(1-x)$,    (f) $x \to 1/(x-2)$,    (g) $x \to 4-6/x$,    (h) $x \to \sqrt{x-2}$.

6. (a) $x \to x+2$,    (b) $x \to 3x$,    (c) $x \to \frac{1}{2}x-2$,
   (d) $x \to \frac{1}{3}x-2$,    (e) $x \to \frac{1}{3}(x-2)$,    (f) $x \to 2x+2$,
   (g) $x \to 2x$,    (h) $x \to \frac{2}{3}(x+2)$,    (i) $x \to \frac{2}{3}(x+6)$,
   (j) $x \to 3x-2$,    (k) $x \to 3(x+2)$,    (l) $x \to 3(x+2)$,
   (m) $x \to x$,    (n) $x \to \frac{1}{3}(x-4)$,    (o) $x \to \frac{2}{3}(x-1)$.

8. (a) Yes,    (b) 5, 36,    (c) 14, 28, 56, 70 …,
   (d) no,    (e) $y$ is a multiple of $x$.

9. (a) 2, 8, 8, 9,    (b) $x$ is a prime number,
   (c) $x = 2$ or 1,    (d) 3, 4, 5, $n+1$.

## Chapter 6

1. $A'(3, -1)$,    $B'(2, -4)$,    $C'(-1, -3)$,    $D'(0, 0)$.

2. (a) $\begin{pmatrix} 3 \\ 2 \end{pmatrix}$,    (b) $\begin{pmatrix} 2 \\ 6 \end{pmatrix}$,    (c) $\begin{pmatrix} 2 \\ 2 \end{pmatrix}$,    (d) $\begin{pmatrix} -3 \\ -2 \end{pmatrix}$;
   (a) and (d) inverse pair.

3. $(-2, 1), (1, 2), (0, -1), (-1, -2)$.

4. $(-1, -1), (-3, 0), (0, 1)$; translation with vector $\begin{pmatrix} -5 \\ 0 \end{pmatrix}$.

5. (a) $\begin{pmatrix} 1 \\ 1 \end{pmatrix}$,     (b) $\begin{pmatrix} 3 \\ -7 \end{pmatrix}$,     (c) $\begin{pmatrix} -2 \\ 3 \end{pmatrix}$,     (d) $\begin{pmatrix} 4 \\ -6 \end{pmatrix}$,

   (e) $\begin{pmatrix} -3 \\ 12 \end{pmatrix}$,     (f) $\begin{pmatrix} 4 \\ -1 \end{pmatrix}$,     (g) $\begin{pmatrix} -7 \\ 18 \end{pmatrix}$,     (h) $\begin{pmatrix} 8 \\ -22 \end{pmatrix}$.

6. 23.3 km, 061°.

7. (a) **e**,     (b) **d**,     (c) **e**+**d**,     (d) 2**d**,
   (e) **e**−**d**,     (f) **d**−2**e**,     (g) 2**d**−2**e**,     (h) **e**−2**d**.

8. $\begin{pmatrix} 2 \\ -3 \\ \frac{1}{2} \end{pmatrix}$,   $\begin{pmatrix} 12 \\ 2 \\ 2\frac{1}{2} \end{pmatrix}$,   $\begin{pmatrix} -6 \\ -2 \\ -2\frac{1}{2} \end{pmatrix}$.

9. (i) **DC**, **EH**, **FG**,     (b) **BH**;
   (ii) (a) **AE**,     (b) **AH**,     (c) **AC**,     (d) **AH**,     (e) **O**;
   (iii) the mid-point of *DH*.


## Chapter 7

1. (a) $(-1, 1), (-1, 3), (-2, 3)$,     (b) $(-1, 1), (-3, 1), (-3, 0)$,
   (c) $(1, -1), (3, -1), (3, -2)$,     (d) $(3, 1), (1, 1), (1, 2)$.

2. $(3·6, 2·7)$.

3. (a) $y = 3$,     (b) $2y + 3x = 2$,
   (c) $6y = 4x + 9$,     (d) $y = 2x - 4$.

4. (a) $(0, 4)$,     (b) $(-3, 0)$,     (c) $(1, -2)$,     (d) $(2, 5)$.

5. $(-2, 4), -143°$.

6. (a) $(-1·2, 0·8), -123°$,     (b) $(1, 1), 180°$,     (c) $(4, 4), 46°$.

7. (a) Rotation of $-45°$ about centre,
   (b) rotation of $180°$ about centre,
   (c) reflection in mediator of $AB$,
   (d) reflection in mediator of $AB$,
   (e) reflection in mediator of $BC$, or rotation of $-45°$ about centre,
   (f) reflection in mediator of $AH$.

8. $A'(3, 1)$,     $B'(3, 2)$,     $C'(5, 2)$; $y + x = 2$;
   $A^*(-1, 1)$,     $B^*(-1, 0)$,     $C^*(-3, 0)$; rotation of $180°$ about $(1, 1)$.

## Chapter 8

1. (a) $(1,2), (1,-2), (5,4)$;
   (b) $(-\frac{1}{2}, \frac{1}{2}), (-\frac{1}{2}, -\frac{1}{2}), (\frac{1}{2}, 1)$;
   (c) $(-3,-2), (-3,2), (-7,-4)$.

2. (a) scale factor 4, centre $(-1\frac{1}{3}, 0)$,
   (b) scale factor $-4$, centre $(4/5, 0)$.

3. $(-1,-1)$; scale factor 2, centre $(-4,-6)$.

4. (a) $(0,1), (-2,2), (0,2), (2,1)$;
   (b) $(2,0), (1,0), (-3,2), (-2,2)$.

7. (a) 7·5 cm,　　　(b) 1300 m,　　　(c) 0·12 km².

8. (a) 68 cm,　　　(b) 232,　　　(c) 380 cm²,　　　(d) 300 m³.


## Chapter 9

1. (a) $(0,0), (-3,0), (-3,-3), (0,-3)$;
   (b) $(0,0), (3,0), (3,1), (0,1)$;
   (c) $(0,0), (2,-1), (3,-1), (1,0)$;
   (d) $(0,0), (4,2), (1,3), (-3,1)$;
   (e) $(0,0), (1,0), (1,-1), (0,-1)$;
   (f) $(0,0), (1,4), (1,5), (0,1)$;
   (g) $(0,0), (-1,1), (2,3), (3,2)$;
   (h) $(0,0), (3,1), (9,3), (6,2)$;
   (i) $(0,0), (1,0), (1,1), (0,1)$.

2. $A(0,3),$　　　$B(-1,2),$　　　$C(2,1)$.

3. Reflection in $y = 0$.

$$\begin{pmatrix} -2 & 4 \\ -1 & 6 \end{pmatrix}, \quad \text{no.}$$

5. (a) $\begin{pmatrix} -1 & 0 \\ 0 & -1 \end{pmatrix}$,　　(b) $\begin{pmatrix} 1 & -1\frac{1}{2} \\ 0 & 1 \end{pmatrix}$,

   (c) $\begin{pmatrix} 1 & 2\frac{1}{2} \\ 1 & 2\frac{1}{2} \end{pmatrix}$,　　(d) $\begin{pmatrix} 1\frac{1}{2} & 0 \\ 0 & -2 \end{pmatrix}$,

   $\begin{pmatrix} \frac{2}{3} & \frac{3}{4} \\ 0 & -\frac{1}{2} \end{pmatrix}$,　　$\begin{pmatrix} -\frac{2}{3} & 0 \\ 0 & \frac{1}{2} \end{pmatrix}$,　　no.

6. $\begin{pmatrix} 2 & -1 \\ -1 & 1 \end{pmatrix}, \begin{pmatrix} 1 & 1 \\ 1 & 2 \end{pmatrix}, \begin{pmatrix} 1 & 1 \\ 4 & -2 \end{pmatrix}, \begin{pmatrix} 2 & 3 \\ 2 & 0 \end{pmatrix}, \begin{pmatrix} -6 & 4\frac{1}{2} \\ 6 & -3\frac{1}{2} \end{pmatrix}$.

8. $A'(3,4),$ $\qquad$ $B'(4,6),$ $\qquad$ $C'(1,2)$; 2 square units, 2.

9. Translation with vector $\begin{pmatrix} a \\ b \end{pmatrix}$; rotation of $90°$ about $(0,0)$ followed by a

translation with vector $\begin{pmatrix} 2 \\ 3 \end{pmatrix}$; $\begin{pmatrix} 0 & -1 \\ 1 & 0 \end{pmatrix}$, $c = 2$, $d = 3$.

## Chapter 10

1. $(1\frac{2}{3}, 3\frac{1}{3})$, $(\frac{3}{4}, 4\frac{1}{4})$, $(9, -4)$.

3. $m = 2$, $c = -1$, $y = 2x - 1$.

4. (a) $2y = x + 7$, $\qquad$ (b) $3y + 5x = 4$, $\qquad$ (c) $y = x - 1$,
   (d) $y + 2x = 6$, $\qquad$ (e) $2y + 3x = 15$, $\qquad$ (f) $4y = 9x + 6$.

7. (a) 3 km, $\qquad$ (b) 2.15 p.m., $\qquad$ (c) 4·7 km.

8. (a) 12 minutes before kick-off, $\qquad$ (b) 2000.

## Chapter 11

1. (a) $x = -2$, $\qquad$ (b) $x = -1\frac{1}{4}$, $\qquad$ (c) $x = -3$, $\qquad$ (d) $x = 6$,
   (e) $x = -\frac{1}{3}$, $\qquad$ (f) $x = 3$, $\qquad$ (g) $x = -2\frac{1}{2}$, $\qquad$ (h) $x = 4$,
   (i) $x = 24$, $\qquad$ (j) $x = -3\frac{1}{2}$, $\qquad$ (k) $x = 5\frac{1}{2}$, $\qquad$ (l) $x = -\frac{1}{2}$,
   (m) $x = -\frac{2}{3}$, $\qquad$ (n) $x = \frac{3}{4}$, $\qquad$ (o) $x = -\frac{1}{2}$, $\qquad$ (p) $x = -3$.

2. (a) $\{x : x > 2\}$, $\qquad$ (b) $\{x : x \geqslant -1\frac{1}{2}\}$, (c) $\{x : x < 1\}$, $\qquad$ (d) $\{x : x \leqslant 1\frac{1}{2}\}$,
   (e) $\{x : x < -9\}$, $\qquad$ (f) $\{x : x > 1\}$, $\qquad$ (g) $\{x : x \geqslant \frac{6}{7}\}$, $\qquad$ (h) $\varnothing$,
   (i) $\{x : x < 1\frac{1}{4}\}$, $\qquad$ (j) $\{$real numbers$\}$.

3. $f : x \to \frac{1}{2}(\frac{x}{3} + 1)$, $\frac{1}{6}(x + 3)$,
   (a) $x = 1\frac{1}{3}$, $\qquad$ (b) $x = \frac{1}{6}$, $\qquad$ (c) $x = 1\frac{1}{6}$,
   (d) $x = 1\frac{2}{3}$, $\qquad$ (e) $x = \frac{3}{4}$, $\qquad$ (f) $x = -1\frac{1}{3}$.

4. (a) $x = 1$ or $x = 2$, $\qquad$ (b) $x = -1$ or $x = \frac{1}{2}$,
   (c) $x = 0$ or $x = -1$, $\qquad$ (d) $x = 3$ or $x = 3$,
   (e) $x = 0$ or $x = -\frac{2}{5}$.

5. (a) $\{x : -1 \leqslant x \leqslant 2\}$, $\qquad$ (b) $\{x : x < -1$ or $x > \frac{1}{2}\}$,
   (c) $\{x : 0 < x < \frac{3}{2}\}$, $\qquad$ (d) $\{x : x \leqslant 0$ or $x \geqslant \frac{2}{3}\}$,
   (e) $\{x : x \neq \frac{1}{2}\}$.

## Chapter 12

1. (a) $x = 1$, $y = 2$, $\qquad$ (b) $x = -2\frac{1}{2}$, $y = 2$,
   (c) $x = 3$, $y = -4$, $\qquad$ (d) $x = \frac{3}{5}$, $y = \frac{2}{5}$,
   (e) $x = \frac{3}{7}$, $y = -\frac{5}{7}$, $\qquad$ (f) $x = 41$, $y = 101$,
   (g) $x = -\frac{3}{11}$, $y = \frac{5}{11}$, $\qquad$ (h) $x = 1$, $y = 5$,

(i) $x = \frac{2}{9}, y = \frac{5}{9},$        (j) $x = 21, y = 19,$

(k) $x = 2\frac{7}{8}, y = -2\frac{1}{8},$      (l) $x = 1, y = 1; x = 2, y = 4,$

(m) $x = 8, y = 1; x = 1\frac{1}{2}, y = 5\frac{1}{3},$     (n) $x = \frac{1}{2}, y = 1\frac{1}{8}.$

2. (a) one, $x = 2, y = 1,$

   (b) infinitely many, $\{(x, y) : x + 2y = 1\},$

   (c) none,          (d) none,

   (e) three, $x = 0, y = 0; x = 1, y = 1; x = -1, y = -1,$

   (f) none.

3. $\begin{pmatrix} 4 & 0 & 0 \\ 0 & 4 & 0 \\ 0 & 0 & 4 \end{pmatrix},$   $x = 2, y = 1, z = -1.$

4. $y = 2x - 3.$         5. $x = 3, y = 2.$

6. 100 cars, 20 coaches.     7. 582 bowls, 205 jugs.

8. Yes, 6.

## Chapter 13

3. (a) 12, 2;       (b) 12, 3.

5. (a) spend 2 hours on hobbies,     (b) spend $2\frac{1}{2}$ hours on homework.

7. 30 detached, 50 semi-detached.

8. (a) $125\pi$ cm$^3$,      (b) 1·4 cm,      (c) 3·3 cm.

## Chapter 14

1. 112°..

2. 4·47 m, 41·8°, 1·48 m.

3. 12·6 cm.

4. (a) 7·27 km,      (b) 16·3 km,      (c) 17·9 km,      (d) 246°.

5. 30·6 m, 34·8°.

6. 2·59 km, 11·5° and 34·5°.

7. (a) 14 m,     (b) 6 m,     (c) 36 s,     (d) 14 s.

8. $\frac{1}{3}.$

9. Between 038·7° and 045°, 042°.

10. (i) (a) $(4\cdot33, 2\cdot5)$,       (ii) (a) $(6\cdot40, 51\cdot3°)$,
    (b) $(-7\cdot06, -3\cdot76)$,         (b) $(4\cdot47, 116\cdot6°)$,
    (c) $(-4\cdot79, 3\cdot61)$,          (c) $(6\cdot71, 243\cdot4°)$,
    (d) $(1\cdot50, -3\cdot71)$,         (d) $(5\cdot83, 301°)$.

11. (a) $\cos\theta = -3/5, \tan\theta = -4/3$,
    (b) $\sin\theta = -0\cdot866, \tan\theta = 1\cdot732$,
    (c) $\sin\theta = 0\cdot814, \cos\theta = 0\cdot581$.

## Chapter 15

1.
| | | |
|---|---|---|
| 2 | 1·75 | 1·7 |
| 0·07 | 0·0736 | 0·1 |
| 2000 | 2370 | 2367(·0) |
| 10 | 11·4 | 11·4 |
| 0·003 | 0·00329 | 0·0 |
| 3000 | 2760 | 2761·9 |
| 0·0003 | 0·000321 | 0·0 |

2.
| | | | |
|---|---|---|---|
| 23 | 10111 | 43 | 32 |
| 59 | 111011 | 214 | 113 |
| 96 | 1100000 | 341 | 165 |
| 49 | 110001 | 144 | 100 |
| 86 | 1010110 | 321 | 152 |
| 3 | 11 | 3 | 3 |
| 125 | 1111101 | 1000 | 236 |
| 182 | 10110110 | 1212 | 350 |
| 5 | 101 | 10 | 5 |

3. (a) $1000101_2$,   (b) $143_7$,            (c) $10212_3$,
  (d) $423_5$,      (e) $144e_{12}$, (e = eleven),   (f) $112_6$,
  (g) $15732_8$,   (h) $11011_2$.

4. (a) 5,     (b) 9,     (c) 4,     (d) 8.

5. (a) (i) $0\cdot11_2$, (ii) $0\cdot9_{12}$,     (b) $11\frac{9}{16}$,
  (c) $0\cdot343_6$,             (d) $32/100$.

6. 6, 24/11/1956.

7. 58 kg, 55 kg, 52 kg.

## Chapter 16

1. (a) $3\cdot7 \times 10^3$,     (b) $2\cdot311 \times 10^4$,     (c) $3\cdot5 \times 10^{-2}$,
  (d) $2\cdot01 \times 10^{-4}$,   (e) $2\cdot25 \times 10$,     (f) $5 \times 10^3$,
  (g) $1\cdot6 \times 10^{-3}$,    (h) $5\cdot6 \times 10^7$,     (i) $4 \times 10^{-3}$.

2. (a) 0·2,  (b) 200,  (c) 20,  (d) 0·02,  (e) 0·6,
   (f) 0·2,  (g) 0·6,  (h) 0·0005,  (i) 10,  (j) 0·000 0001.

3. 28·2 cm, 21·15 cm.

4. Between 96·525 km and 103·525 km.

5. 5·0625 cm² and 4·6225 cm²; one.

## Chapter 17

1. (a) 4,  (b) 2·32,  (c) 4·15, 96, 3·3.

2. (a) 0·726,  (b) 2·581,  (c) $\bar{1}$·559,  (d) 3·575,
   (e) $\bar{4}$·724,  (f) 4·993,  (g) $\bar{6}$·865,  (h) 6·344.

3. (a) 2·59,  (b) 520,  (c) 0·270,  (d) 0·0684,
   (e) 3600,  (f) 0·000109,  (g) 17300,  (h) $10^5$.

4. (a) 223,  (b) 26·2,  (c) 198,  (d) 6670,
   (e) 6·17,  (f) 0·187,  (g) 3350,  (h) 0·00275,
   (i) 60·0,  (j) 1·15.

5. 36·8 cm³, 129 g.    6. 6·36.

7. (a) 3,  (b) 2,  (c) −2,  (d) 10,
   (e) 0,  (f) 4,  (g) 0·5,  (h) 2.

## Chapter 18

1. 19·7.    2. 6·42.    3. 1·92.
4. 0·0884.    5. 19·3.    6. 0·848.
7. 0·104.    8. 0·325.    9. 59·0.
10. 1·83.    11. 1·17.    12. 45·6.
13. 56·4.    14. 76·8.    15. 6·89.
16. 24 cm².    17. 1·02.    18. 35·4.

## Chapter 19

1. (a) $6x^2 + 7x + 2$,  (b) $15x^2 - 19x + 6$,
   (c) $16x^2 - 25$,  (d) $-12x^2 + 2x + 2$,
   (e) $2x^2 + 5xy + 2y^2$,  (f) $12t^2 + 5t - 28$,
   (g) $8p^2 - 22p + 12$,  (h) $9a^2 + 27ab + 20b^2$,
   (i) $6y^3 - 7y^2 + 11y - 6$,  (j) $ab + ad + cb + cd$.

166

2. (a) $t = \dfrac{v-u}{f}$,　(b) $x = \pm\sqrt{\dfrac{y-b}{a}}$,

   (c) $x = \pm\sqrt{\dfrac{a^2w^2 - v^2}{a}}$,　(d) $t = \dfrac{l - l_0}{al_0}$,

   (e) $a = \dfrac{b^2 - D}{4c}$,　(f) $l = \dfrac{t^2 g}{4\pi^2}$,

   (g) $r = \dfrac{2wa}{w - av^2}$,　(h) $l = \dfrac{r(r - 2a)}{a}$.

3. $C = 100 - ax$, $C = yb$, $yb + ax = 100$.

4. $v = \pm\sqrt{(u^2 + 2fs)}$.

5. (a) $a = \dfrac{P}{6 + \pi}$,　(b) $a = \sqrt{\dfrac{2A}{8 + \pi}}$.

6. $\sqrt{(8n)}$, $2n$ is a perfect square.

7. $x^2 - 4x - 5$, $x = \pm\sqrt{(y + 9)} + 2$;　(a) $x = 5$ or $x = -1$,
   (b) $x = 5{\cdot}46$ or $x = -1{\cdot}46$,　(c) $x = 3{\cdot}41$ or $x = 0{\cdot}59$,
   (d) no solutions,　(e) $x = 3{\cdot}24$ or $x = -1{\cdot}24$.

## Chapter 20

1. (a) 2　7　8　20,　(b) $-17\frac{1}{2}$　$-3\frac{1}{2}$　0　$11\frac{3}{8}$,
   (c) $10\frac{2}{3}$　$\frac{1}{6}$　6　$\frac{2}{3}$,　(d) 21　14　3　2,
   (e) 1　4　400　$11\frac{1}{9}$,　(f) 0　2　5　5·64.

2. (a) 75,　(b) $1\frac{3}{5}$,　(c) $\frac{1}{4}$.

3. (a) $y \propto x$, $y = \frac{2}{3}x$,　(b) $y \propto x$, $y = 0{\cdot}75x$,　(c) $y \propto 1/x$, $y = 22{\cdot}5/x$.

4. 8·8.　　5. $y \doteqdot 0{\cdot}74x^2$.

6. 3·6　　7. £25, 8p, $a = 0{\cdot}08$, $b = 25$.

8. 3.　　9. 6·3.

## Chapter 21

1. $-1$; 7; $4\frac{1}{2}$ sq. units.

2. (a) 84,　(b) 21·7,　(c) 62·7,　(d) 192.

3. 6; 12; 30; 42; 3, 7, 11 …

4. 712·5 m.

5. 33 km; 66 km/h.

6. (a) 106°C; 48°C,　(b) 48 sec,　(c) 126°C/min; 7°C/min.

7. (a) 810 m; 3 hours after starting,
   (b) after 2 hours; no,
   (c) 150 m/h.

8. (a) 580 cars/h,      (b) 2110 cars.

9. 30; 19.

10. (a) 5·8 m/s,      (b) 4·8 m/s,      (c) 11·5 m/s.


## Chapter 22

2. (a) $\dfrac{1}{13}$,      (b) $\dfrac{1}{4}$,      (c) $\dfrac{1}{52}$,      (d) $\dfrac{3}{13}$,      (e) $\dfrac{9}{13}$.

3. (a) $\dfrac{5}{8}$,      (b) $\dfrac{3}{8}$;   $\dfrac{15}{28}$.

4. (a) $\dfrac{1}{8}$,      (b) $\dfrac{3}{8}$,      (c) $\dfrac{2}{5}$,      (d) 1,      (e) 0.

5. (a) $\dfrac{6}{125}$,      (b) $\dfrac{17}{500}$,      (c) $\dfrac{459}{500}$.

6. (a) $\dfrac{17}{150}$,      (b) $\dfrac{23}{150}$,      (c) $\dfrac{53}{150}$.

7. (a) 0,      (b) $\dfrac{1}{6}$,      (c) $\dfrac{15}{36}$,      (d) $\dfrac{11}{36}$.

8. $\dfrac{1}{9}$.

9. (a) $\dfrac{27}{80}$,      (b) $\dfrac{9}{40}$.

10. (a) $\dfrac{1}{16}$,      (b) $\dfrac{17}{24}$,      (c) $\dfrac{1}{24}$,      (d) $\dfrac{9}{48}$.

11. $\dfrac{1}{5}$.


## Chapter 23

1. (a) 43,      (b) $48\frac{6}{11}$,      (c) 43.

3. (a) 101,      (b) 101,      (c) each member,      (d) 12.

4. (a) £1700,      (b) £600.

5. 12·1, 12·1, no.

6. 605·5 m.

7. 60, 60, 62·5; 56·7, 65, 55.

8. A 100–74,    B 74–65,    C 65–49,    D 49–33,    E 33–0.

## Chapter 24

1. 211·4 m²,      9·42 m².

2. $1·1 \times 10^{-3}$ cm.

3. 3·96 m³,      22 200 kg.

4. 4 cm.

5. 49·5.

6. (a) 12·5,      (b) 36·5,      (c) 42,      (d) 59·5.

7. 27 : 19 : 7 : 1.

## Chapter 25

1.                      22·6 cm     31·4 cm²
                       51·2 cm     141 cm²
                       8·39 cm     4·23 cm²
                 95·5°             30 cm²
     9·77 cm         40·0 cm

2. 570.

3. 12 m × 24 m, 62 m².

4. (a) 378 n. miles,      (b) 3160 n. miles,      (c) 3068 n. miles.

5. (a) 2320 n. miles,      (b) 8000 n. miles,      (c) 3980 n. miles.

6. (a) 10234 n. miles,      (b) 6922 n. miles.

7. 31° 51′ W, 29° S; 31° 51′ W, 42° 20′ S.

8. 8·66 cm, 12·2 cm.

10. 2° 30′ W.

## Chapter 26

1. (b), (c) and (f); (a) and (e).

3. (a) $\begin{pmatrix} 2 & 1 & 2 \\ 1 & 0 & 0 \\ 2 & 0 & 0 \end{pmatrix}$, (b) $\begin{pmatrix} 0 & 2 & 0 & 0 \\ 2 & 0 & 1 & 1 \\ 0 & 1 & 0 & 1 \\ 0 & 1 & 1 & 0 \end{pmatrix}$, (c) $\begin{pmatrix} 0 & 1 & 0 & 1 \\ 0 & 0 & 0 & 1 \\ 1 & 0 & 1 & 0 \\ 1 & 0 & 0 & 0 \end{pmatrix}$.

6. (a) $\begin{pmatrix} 1 & 0 & 1 & 1 & 0 \\ 1 & 1 & 0 & 0 & 1 \\ 0 & 1 & 1 & 0 & 0 \\ 0 & 0 & 0 & 1 & 1 \end{pmatrix}$, (b) $\begin{pmatrix} 1 & 0 & 1 \\ 1 & 1 & 0 \\ 1 & 1 & 0 \\ 0 & 1 & 1 \\ 0 & 1 & 1 \end{pmatrix}$, (c) $\begin{pmatrix} 1 & 1 & 1 & 0 \\ 1 & 1 & 1 & 1 \\ 1 & 1 & 0 & 1 \end{pmatrix}$,

(d) $\begin{pmatrix} 0 & 1 & 1 & 1 \\ 1 & 0 & 1 & 1 \\ 1 & 1 & 0 & 0 \\ 1 & 1 & 0 & 0 \end{pmatrix}$, $\begin{pmatrix} 3 & 1 & 1 & 1 \\ 1 & 3 & 1 & 1 \\ 1 & 1 & 2 & 0 \\ 1 & 1 & 0 & 2 \end{pmatrix}$, $\begin{pmatrix} 2 & 2 & 2 \\ 2 & 2 & 2 \\ 2 & 2 & 0 \\ 0 & 2 & 2 \end{pmatrix}$.

7. $\begin{pmatrix} 0 & 1 & 1 & 2 \\ 1 & 0 & 0 & 0 \\ 1 & 0 & 0 & 1 \\ 2 & 0 & 1 & 2 \end{pmatrix}$, $\begin{pmatrix} 6 & 0 & 2 & 5 \\ 0 & 1 & 1 & 2 \\ 2 & 1 & 2 & 4 \\ 5 & 2 & 4 & 9 \end{pmatrix}$;

(a) 5, (b) 9, (c) 9.

8. (a) No, (b) yes.

9. $\begin{pmatrix} 1 & 1 & 1 \\ 1 & 1 & 1 \\ 0 & 1 & 1 \\ 0 & 0 & 1 \end{pmatrix}$, $\begin{pmatrix} 1 & 1 & 1 \\ 1 & 1 & 1 \\ 0 & 1 & 1 \\ 0 & 0 & 1 \end{pmatrix}$, no, no.

## Chapter 27

1. (a) Line symmetry about one diagonal.

(b) Rotational symmetry of order 4.

(c) Point symmetry.

(d) Line symmetry about mediator of non-arrowed side.

3. $(3, 1)$, $\begin{pmatrix} 3 & 1 \\ 1 & 3 \end{pmatrix}$, line symmetry about $y = x$ and $x + y = 4$, point symmetry about $(2, 2)$.

4. 2 straight lines parallel to, and equidistant from, $AB$.

6. $\mathbf{AD} = \mathbf{BC} = \mathbf{x} + \mathbf{y}$.

7. $62°$.

8. $OT$, $90°$, $2\theta$.

10. Straight line.

11. In direction $P$ to $R$, no, $\mathbf{PX} - \mathbf{XR} = \mathbf{QX} - \mathbf{XS} = \mathbf{O}$, diagonals bisect each other.

## Chapter 28

1. (a) Closed under ×, − and +.　　(b) Closed under × and +.

　(c) Closed under ×.　　　　　　(d) Closed under ∪ and ∩.

　(e) Closed under multiplication mod. 10.

2. Y = reflection in $y = 0$, I = identity, rectangle or rhombus.

3. (a)

| × | 1 | 2 | 3 | 4 |
|---|---|---|---|---|
| 1 | 1 | 2 | 3 | 4 |
| 2 | 2 | 4 | 1 | 3 |
| 3 | 3 | 1 | 4 | 2 |
| 4 | 4 | 3 | 2 | 1 |

(b)

| + | 0 | 1 | 2 | 3 |
|---|---|---|---|---|
| 0 | 0 | 1 | 2 | 3 |
| 1 | 1 | 2 | 3 | 0 |
| 2 | 2 | 3 | 0 | 1 |
| 3 | 3 | 0 | 1 | 2 |

(c)

| × | 0 | 1 | 2 | 3 |
|---|---|---|---|---|
| 0 | 0 | 0 | 0 | 0 |
| 1 | 0 | 1 | 2 | 3 |
| 2 | 0 | 2 | 0 | 2 |
| 3 | 0 | 3 | 2 | 1 |

(d)

| | I | X | Y | R |
|---|---|---|---|---|
| I | I | X | Y | R |
| X | X | I | R | Y |
| Y | Y | R | I | X |
| R | R | Y | X | I |

　(i) (a) 1,　　(b) 0,　　(c) 1,　　(d) I.

　(ii) (a) (3, 2) inverse pair, 1 and 4 self-inverse.

　　(b) (3, 1) inverse pair, 0 and 2 self-inverse.

　　(c) 1 and 3 self-inverse.

　　(d) All elements self-inverse.

4. (i) Commutative and associative.

　(ii) Not commutative, not associative.

　(iii) Commutative, not associative.

5. (a) $p+q$,　　(b) $p.q$,　　(c) $p+q$,　　(d) 0.

6. (a) $(a+b).a = a$,

　(b) $(a+b+c).(a'+b)+c = b+c$,

　(c) $(a+b).(b'+c')+c.a = a+bc'$.

7. (a) I use my car and travel by bus, or I am sometimes late.

　(b) I don't travel by bus but use my car.

(c) I use my car, or I travel by bus and am sometimes late.

(d) I use my car and am never late or I travel by bus and am sometimes late.

8. (a) $c \wedge a \wedge b$,　　　(b) $\sim c \wedge b$,

(c) $\sim(a \vee b) \wedge c$,　　　(d) $a \wedge (\sim c \wedge \sim b)$ or $a \wedge \sim(c \vee b)$.

# Miscellaneous Exercises

## Exercise A

1. $(-3, -8)$.

2. $3\frac{1}{3}$,　$\begin{pmatrix} 1\frac{1}{4} & \frac{1}{2} \\ -\frac{3}{4} & \frac{1}{2} \end{pmatrix}$.

3. $x = \frac{3}{5}, y = -\frac{2}{5}$.

4. 17, 3.

5. $1\frac{1}{4}$.

6. (a) $27\frac{7}{9}$,　　(b) 24.

7. $2 \times 10^5$.

8. $m$ is even.

9. $900°$, $128\frac{4}{7}°$.

## Exercise B

1. 2, 3, 19.

2. $\begin{pmatrix} -3 \\ 5 \end{pmatrix}$.

3. $1\frac{1}{2}$, yes.

4. $x = -3$, $\{x : x > -3\}$.

5. $x = 3\cdot7, y = 1\cdot3$.

6. 1/12.

7. $3\cdot16, 2\cdot73$, not associative.

8. $\begin{pmatrix} 2 & 1 & 0 \\ 1 & 0 & 2 \\ 0 & 2 & 0 \end{pmatrix}$,　$B$ and $C$.

9. $192 \text{ cm}^2$.

10. $t = \sqrt{\dfrac{2s}{g}}, t = 1\frac{5}{7}$.

## Exercise C

2. $16° 40'$, $21° 42'$.

3. $\begin{pmatrix} 1 & 2 \\ 0 & 1 \end{pmatrix}$.

4. 2.20 p.m.

5. 30.

6. 422 kg.

7. $39\cdot3$ cm.

8. 2/15.

9. 3.

10. (a) yes,　　(b) no,　　(c) $\mathbf{A}$ is singular.

## Exercise D

1. True.
2. False, $\{x:x^2-9=0\}\cap\{x:x<0\}=\{-3\}$.
3. True.
4. False, the inverse of $fg$ is $g^{-1}f^{-1}$.
5. True.
6. True.
7. False, factor of 4.
8. False, infinite number of solutions.
9. False, 7100.
10. True.
11. True.
12. False, probability is $\frac{3}{8}$.
13. True.
14. False (latitude $40°$ N, longitude $50°$ E).
15. False, 0 has no inverse.

## Exercise E

1. (b).
2. (d).
3. (a).
4. (b).
5. (b).
6. (a).
7. (d).
8. (b).
9. (b).
10. (d).

## Exercise F

1. (d).
2. (b).
3. (c).
4. (a).
5. (c).
6. (c).
7. (b).
8. (b).
9. (d).
10. (b).

## Exercise G

2. (a) $\frac{3}{8}$,
(b) $\frac{15}{28}$,
(c) $\frac{3}{8}$,
(d) $\frac{5}{8}$.
3. (a) 8 s,
(b) 80 m,
(c) 20 m/s.
6. (a) $\frac{7}{9}$,
(b) $\frac{3}{11}$,
(c) $\frac{280}{333}$.
7. $10\frac{5}{7}\%$.
8. $232\ \text{cm}^3$.
9. $97\cdot5°$.
10. $(3,-3\frac{1}{2})$, $(3,1\frac{2}{3})$, $(\frac{20}{17},-\frac{13}{17})$.

## Exercise H

1. $\begin{pmatrix} 6 \\ -3 \end{pmatrix}$, $\quad$ **OM** = **OA** $+\frac{1}{2}$**AB**, $\quad$ $M$ is $(4,3\frac{1}{2})$, $\quad$ $N$ is $(3,4)$.

173

2. $22 \cdot 6°$, $1 \cdot 15$ m.                                    3. $5/32$.

4. (a) $1650$ cm$^2$,          (b) $2150$ cm$^2$,          (c) $800$ cm$^2$.

5. $a = 3 \cdot 7$,      $b = -0 \cdot 4$.                6. $\begin{pmatrix} 1 & -4 \\ 0 & 1 \end{pmatrix}$.

7. $11, 4$.                                          8. $\varnothing, 4$.

9. $m, m = (y - c)/x$.

10. (a) one,          (b) none,          (c) one,          (d) none.

## Exercise I

1. $17\%$.                    2. $3x - 2y = 0$.                3. $3, 4, 12$; $a, c, ac$.

4. (a) $4$,                    (b) $19$,                    (c) $5$.

5. (a) $\{2x\}$,                    (b) $\{-1\frac{1}{3}, 2\}$,                (c) $\{-\sqrt{5}, -1\frac{1}{3}, 2, \sqrt{5}\}$.

6. $26 \cdot 3$ km, $347 \cdot 6°$.        7. $2 \cdot 84$ m.                8. $10$p, $15$p, $20$p.